AF616479

THE SHALE OIL AND GAS DEBATE

Philippe CHARLEZ
Pascal BAYLOCQ

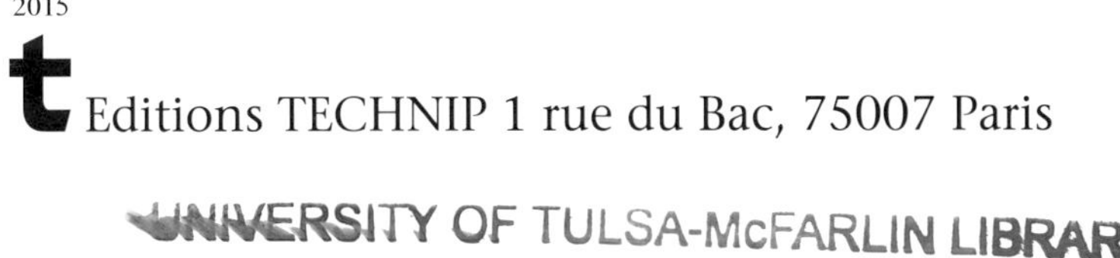
2015
Editions TECHNIP 1 rue du Bac, 75007 Paris

UNIVERSITY OF TULSA-McFARLIN LIBRARY

FROM THE SAME PUBLISHER

- Our energy future is not set in stone.
  P. CHARLEZ
- After the US Shale Gas Revolution.
  T. BROS
- The Oil & Gas Engineering Guide.
  H. BARON
- Sharing the oil rent – Current situation and good practice.
  G. DARMOIS
- The Geopolitics of Energy.
  J.-P. FAVENNEC

Senior editor: Alexandra Lepinay
Editorial assistant and page layout: Sarah Funel
Cover: Cécile Hébrard
Designer: Interscript

Translated and english proof by Cathy Brewerton (www.anglofile.fr)
Cover image: mixed media by Marie Lise Charlez entitled « Migration »
www.marielisecharlez.com

All rights reserved. No part of this publication may be reproduced or transmitted in any form or by any means, electronic or mechanical, including photocopy, recording, or any information storage and retrieval system, without the prior written permission of the publisher.

© Éditions Technip, Paris, 2015
Printed in France

ISBN 978-2-7108-1153-4

TN858
.C5313
2015

Foreword

Among other obstacles, the energy policy in our country suffers from two serious handicaps, the consequences of which could prove extremely costly and crippling for its economy. The first is a result of the deep-seated contradictions between the necessary and admirable ambitions of energy transition in the long-term and the imperatives of the economy in the short to medium term. The second is the systematic and unjustified denigration of the oil and gas industry, preventing over the last few years a calm and serious discussion of the subject of source rock hydrocarbons, hastily dubbed *"Shale oil and gas"*, to evaluate its potential and advantages.

The main cause of both handicaps is an exaggerated politicization of the ecology and an ideological dogmatism which categorically refuses any fallback on fossil fuels without even stopping to evaluate the positive impacts that it could have. This political and dogmatic version of ecology has spawned a plethora of absurdities and technical inaccuracies, leading up to the vote in favor of the law of July 23, 2011, which prohibits the use of hydraulic fracturing, thereby condemning any possible development of shale oil and gas in France. In the wake of this series of imprecisions and technical falsehoods, the oil and gas industry had to react.

For this reason, in 2011 the GEP-AFTP* created a think tank on Hydrocarbon Source Rocks, the main purpose of which is to establish the truth based on undisputable technical and scientific arguments - a truth which will explain why many European countries have decided to embark on the exploration of these resources which could help them improve their economic performance. Remember that French oil and gas imports represent 90% of the national trade deficit.

This book, backed by the GEP-AFTP, was authored by Philippe Charlez and Pascal Baylocq. Written for a non-specialist readership, it explains in 20 simple Questions and Answers all the geopolitical, technical and societal problems that have arisen during the course of the controversy. It is a complement to the film *"Shale Gas"* produced by the GEP-AFTP in early 2014, which denounced among other falsehoods the fraudulent image of the faucet on fire from the film *Gasland.*

Jean Ropers
President of GEP-AFTP

Preface

In June 2010, a delegation from the French Academy of Technologies flew to Washington to meet the main players in US energy. Their message was clear – the priority for the United States is its energy independence, which relies on two mainstays: offshore oil and shale gas that we were in the process of discovering. It was at the time of the Macondo disaster in the Gulf of Mexico. Naturally, the media spotlight was focused more on this accident than on shale gases, but the revolution was already on the horizon.

In the space of a few years, the United States have become the world's leading gas producers ahead of Russia and the production forecasts are increasing every year. Gas prices plummeted to US$ 3/MBTU (equivalent to US$ 20/boe) and even to a low of US$ 2/MBTU in April 2012.

It is a real revolution with many different consequences. Since 2011 electricity, which had been produced principally by coal-fired gas stations in the past, is now chiefly produced by gas-fired power stations.

Though the coal lobby is particularly forceful in the United States, there are fewer buyers to be found within the national territory. Coal has become abundant and its prices have fallen by 30%. Consequently it is massively exported, in particular to Europe, where the situation is the reverse: it is cheaper to produce electricity from coal than from gas, causing the shutdown of modern gas-fired power stations.

This contrast is the result of the disparity in the market systems because, although there is a world coal and oil market across the globe, there is not really a gas market to speak of.

Gas is exchanged across the world essentially through gas pipelines or in its liquid form (LNG – Liquefied Natural Gas) by LNG carriers. They are governed by long-term bilateral contracts[1] which depend on the average prices of oil products[2]. However, short-term prices of gas consumed on site, or spot prices, fluctuate according to supply and demand. This is why it has been possible to observe simultaneously extremely low prices[3] in the United States where spot prices depend on supply and demand, and high prices in Europe or extremely high prices in Japan where they result from long-term contracts and are essentially indexed on oil. As long as the United States does not export the gas it produces and continues to massively export coal, the price of coal will be low and gas prices outside North America will remain virtually the same. World gas contracts can be modified only by a change in the balance of power triggered by the production of new gas resources, particularly shale gas, in gas importing countries. For this reason, we must evaluate our potential reserves in source rock gas, as it is the only way to ensure a solid negotiating position.

A hot debate has broken out in the United States between the gas industry who wants to export with high margins, and the American consumers and chemical industry (where methane is used as a raw material) who want to preserve this essential competitive edge. Four liquefaction units have already received the necessary legal authorizations (Sabine Pass, Freeport, Lake Charles and Cove Point) in the United States.

The future production of shale oil and gas worldwide is therefore a factor likely to modify the current energy balance. Even if the impact on prices will be limited to liquid hydrocarbons, gas producers and exporters are likely to see their sale prices decrease. This is particularly the case for Gazprom and Russia, who have significantly reduced their activities in the Arctic zone (e.g. the Shtokman project in the Barents Sea has been postponed) owing to excessive exploitation costs in the new gas-dominant

1. Contracts are often signed for durations of 10 to 20 years with a view to prices increasing rather than decreasing.
2. In the 1970s, the price of gas did not have an independent value on the markets, it was indexed on oil prices.
3. Gas prices were abnormally low in the United States, dropping to between 2 and 3 US$/MBTU in the last few years. This stems from the fact that in so-called *"tight oil wells"*, shale oil and shale gas are produced simultaneously. Oil is sold at market prices and allows for high financial margins, whereas gas is simply a by-product. The internal market in the United States has now become organized and the price of gas stands at around US$ 4/MBTU.

context. As for the events in Ukraine, these should have very little impact on gas prices. Coal exporters seem less exposed than gas exporters even though new mines have been opened in preparation for the expected increase in demand from China and India, where electricity generation is essentially based on coal. Prices have probably already dropped. In conclusion, the American coal producers and the Russian gas producers are powerful opponents, resolved to the exploitation of shale gases.

In 2011, the French discovered shale oil and gas, mainly after seeing the *Gasland* film. The documentary made its name by putting forward the argument of the faucet on fire, an image which has since been proved as a deliberate attempt to fan the flames of shale gas opponents[4]. Yet the damage was done, and public opinion saw unconventional hydrocarbons not as a possible opportunity for our industry and our trade balance, but simply as a major threat to our environment.

Why did nobody think it was important to talk about new unconventional hydrocarbons before the Gasland scandal? Already in 2009, permits had been registered and mining titles granted following investigation by the State services and local reports. Upon our return from the US in summer 2010, nobody paid the slightest attention to what we were saying, and it wasn't for want of trying. By the law of July 13, 2011, the government prohibited hydraulic fracturing. It is worth noting that this law was voted despite the fact that the voters who claimed to be in the parliamentary majority at the time were not really against the exploitation of shale gases and in favor of exploration initiatives; the opinion of those who claim to be in the current parliamentary majority is now generally negative.

In 2011, the Academy of Technologies' commission on Energy and Climate Change published a recommendation dated July 2011 and which had been the subject of a press release on May 24, 2011. It proposed a determined process of research and demonstration founded on the existence of resources and considering the ecological acceptability of the technologies involved. The publication of this recommendation in the press or in public meetings stirred up two types of reactions.

4. http://vimeo.com/user15547788/review/86237050/77a47b7662

Proving the existence of resources in no way implies a fast-track program for their exploitation. While the international gas supply is cheap and abundant, we can leave this resource which, for once, will not be a debt, for future generations. However, one day at a public meeting an old lady said to me: *"if exploiting shale gases is harmful to the environment, then the gift is a Trojan horse, it's worse than a debt"*. I answered her saying that I could not believe that we would not be able to produce these oils and gases in a way that was 'green and clean' and I am more and more convinced of the fact. She was right though in thinking that time it would take time to find the right solutions.

Another line of criticism comes from the partisans of intermittent renewable energies who fear that the efforts invested in developing these resources will reduce in equal measure the efforts required to develop wind and solar power. In France we do not need to produce electricity using gas, or using fuel oil for that matter, except for a few exceptional peak times when the winter is extremely cold. The main consumers of natural gas are the chemical industry and heating installations and the new French thermal regulation RT 2012 actually encourages the use of gas. The production of oil and gas in France will simply enable us to reduce our imports, but will have no impact whatsoever on our greenhouse gas emissions. It would be more useful to focus the debate on the similarities and differences between the exploration and exploitation of shale gas and the large scale development of biogas.

In writing The shale oil and gas debate as part of the activities of the Work Group on Source Rock Hydrocarbons from the GEP-AFTP[5], Philippe Charlez and Pascal Baylocq had input and opinions from the best in the field. It is a didactic work by virtue of its presentation, vocabulary and intellectual approach. It should shed light on the issue for all those who are seeking a clear idea of the bigger picture before the major climate event which is to take place in Paris in 2015.

Bernard Tardieu
Président de la Commission Énergie et Changement Climatique
Académie des Technologies

5. *Groupement des Entreprise et des Professionnels des Hydrocarbures et des Énergies Connexes* – Group of French oil companies and related industries

Abbreviations and units

AFTP	Asssociation Française des techniciens du Pétrole
BP	British Petroleum
CAPEX	Capital Expenditures
CLAR	Comité de Liaisons, d'Actions et de Réflexion
EIA	Energy Information Association
EUR	Estimated Ultimated Recovery
GEP	Groupement des Entreprises Pétrolières
GHG	Green House Gas
IEA	International Energy Agency
IOC	International Oil Company
IRR	Internal Rate Return
LNG	Liquefied Natural Gas
LPG	Liquid Petroleum Gas
NORM	Naturally Occurring Radioactive Material
NPV	Net Present Value
OPEX	Operational Expenditures
PWRI	Produced Water Re-Injection
SRV	Stimulated Rock Volume
WTI	West Texas Intermediate

Multipliers			
Pico	10^{-12}	p	
Nano	10^{-9}	n	
Micro	10^{-6}	μ	
Milli	10^{-3}	m	
Unité	10^{0}		
Kilo	10^{3}	k	Thousands
Mega	10^{6}	M	Millions
Giga	10^{9}	G	Billions
Terra	10^{12}	T	Trillions

Symbol	Unit
bbl	Barrel Oil
boe	Barrel Oil Equivalent
toe	Ton Oil Equivalent
bbl/day	Barrel per day
bep/day	Barrel Oil Equivalent per day
cf	Cubic feet
BTU	British Thermal Unit
MWh	MegaWatt hour

Conversion factor	
1 boe = 0,14 toe	1 toe = 7,14 boe
1 boe = 5,8 MBTU	1MBTU = 0,17 boe
1 boe = 5,8 kcf	1 kcf = 0,17 boe
1 boe = 1,63 MWh	1 MWh = 0,61 boe
1 kcf = 1 MBTU 1 kcf = 0,28 MWh 1 MWh = 3,5 MBTU	

Table of contents

APPENDICES

The American shale gas and oil revolution: Why? How? Where?

For the United States, the period from 1970-2006 begins and ends with two key dates: a peak oil production in 1970 at 11,3 Mboepd and a peak in oil imports at 14 Mboepd in 2006, a situation that was no longer tenable for the leading world economy. In 2013, US oil production increased by 38% and imports dropped by 43% compared with 2006. In the same period, gas production rose by 27% and imports fell dramatically by 60 %. In six years, the United States reduced its oil dependency from 67% to 47% and became almost self-sufficient regarding gas supplies. This unexpected earthquake that shook the oil & gas landscape is called "*the shale oil and gas revolution.*"

US energy dependency alone does not explain this revolution. Technically speaking, there is nothing revolutionary as it is the result of two mature technologies – horizontal drilling and hydraulic fracturing. Yes, the economic context was favorable, with the barrel of oil hovering around the US$ 150 mark and that of gas nearing US$ 10/MBTU. Yet the shale oil and gas revolution is not due to these exogenous factors alone. It is also upheld by four endogenous pillars: (1) an extremely detailed knowledge of the subsurface; (2) an oil and gas culture with full backing from the political system; (3) highly favorable mining legislation in that the subsurface belongs to the landowner (4) a near monopoly regarding world oil equipment with, for example, 2,000 drilling rigs operating in the US out of the 2,400 worldwide. As far as gas is concerned, a fifth factor comes into play in the form of a very dense gas-gathering network which means that gas can be easily distributed across the entire country. The four endogenous factors mentioned above fostered the implementation of factory developments, with impressive operational (drilling time) and economic (well costs) performances. Exporting this model to contexts in which the cost of a well is much higher is not necessarily a viable option.

Although there are many play distributed across the US continent, the five largest together represent 65% of oil production and 66% of gas production. For oil, the largest are Bakken (north Dakota) and Eagle Ford (southern Texas) and for gas, Marcellus (Pennsylvania), Barnett (northern Texas) and Haynesville (Texas and Louisiana).

Why?

The United States relationship with oil has always been somewhat "*incestuous*". Most of the major historical events in the second half of the XXth century were underpinned by the obsession with American energy security, a chromosome woven into the genetic heritage of American citizens. For Americans, having fuel cheap enough to allow them to travel at will is almost a constitutional right. If we focus on the years 1970-2006 (**Figure 1**) the key events during the course of this period are almost all directly or indirectly oil-related: the two oil crisis in 1973 and 1979, the counter shock in 1986, the ensuing collapse of the Soviet Union and the two Gulf wars. As if by chance, they are bracketed by two less well-known dates. In 1970, the United States reached its oil peak at 11,3 Mboe/day and in 2006, its crude oil imports topped 14 Mboe/day whereas production fell to less than 7 Mboe/day[1]. Though a lot less notable than its oil dependency, the United States gas dependency also rose during the same period and the country had not been self-sufficient since the end of the 80s. An almost intolerable situation for the leading economic power, which had done so much to ensure its energy independence. Its final massive attempt, under the banner of "*Iraqi freedom*[2]" was not a resounding success.

Let's take our analysis forward now to the following six years, i.e. the period between 2006 and 2013. For the first time since 1970, US oil production increased by 38% and at the same time, its imports decreased by 43%. Natural gas production also increased by 27% over the same period and gas imports fell dramatically by 60% (**Figure 2**). In six years, the United States reduced its oil dependency by 20% (from 67% in 2006

1. World consumption in 2013 was close to 90 million barrels a day.
2. The name of the military operation launched by George W. Bush, also known as "the Second Gulf War".

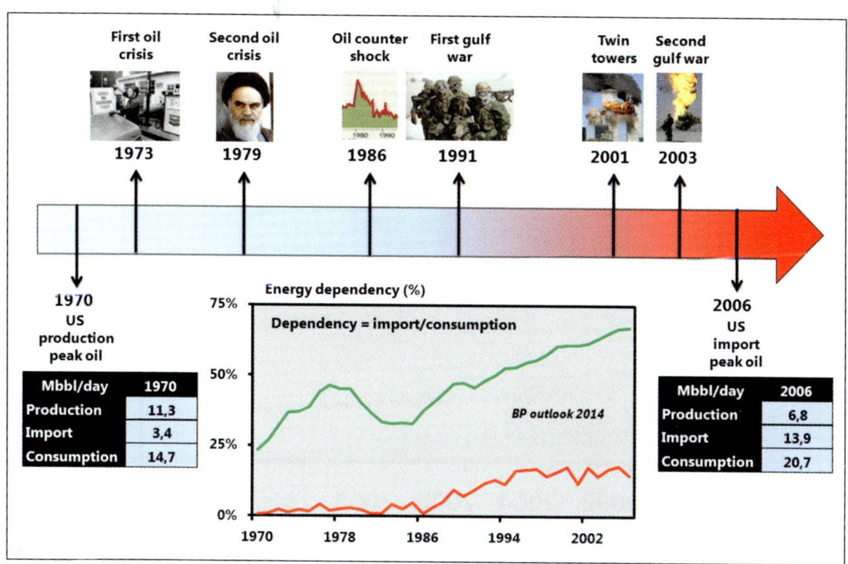

Figure 1 – 1970 – 2006 many oil-related events occurred between the US production and import peaks (BP 2014 Outlook).

to 47% in 2013) and became almost self-sufficient again regarding gas supplies (5% of remaining gas dependency). Naturally, the crisis in 2008 marked the collective mind and Americans became a little more reasonable regarding their excessive consumption of energy. But something else happened too… something that nobody saw coming – a true revolution, the shale oil and gas revolution. The 'explosion' was as sudden as it was unexpected. Between 2006 and mid 2014, shale gas production multiplied by 17, rising from less than 0.35 Mboe/day to 6 Mboe/day[3] and that of shale oil multiplied by ten, from 0.3 Mboe/day to 3 Mboe/day, making the United States the world's leading source of growth, well ahead of Saudi Arabia and Russia. The phenomenon conjured up memories of the gold rush in the mid XIXth century. IOCs hurried to buy out independent companies[4], nuclear projects were put on standby, renewable energies suffered[5], the price of gas and coal plummeted and hundreds of thousands of jobs were created.

3. i.e. an increase equivalent to French annual national consumption.
4. EXXON acquires XTO for $ 41G - http://www.lefigaro.fr/societes/2009/12/14/04015-20091214ARTFIG00527-exxon-rachete-xto-energy-pour-41-milliards-de-dollars-.php
5. http://lecercle.lesechos.fr/economie-societe/energies-environnement/221134593/gaz-schiste-opportunite-manquee

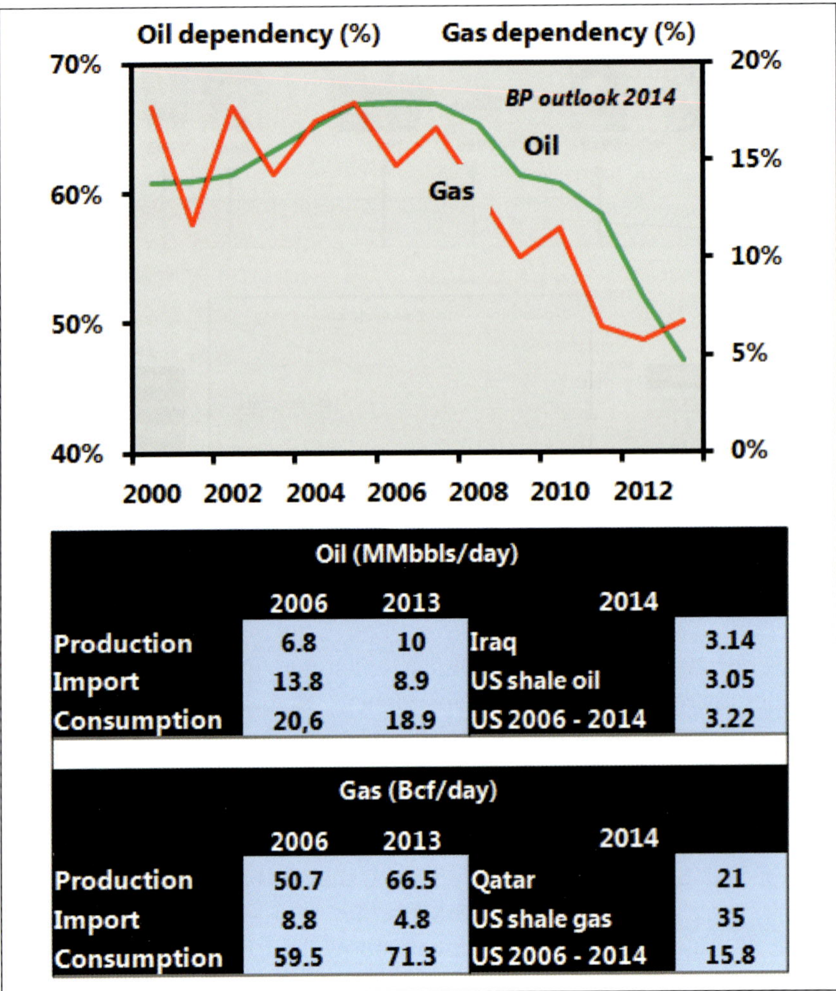

Oil (MMbbls/day)				
	2006	2013	2014	
Production	6.8	10	Iraq	3.14
Import	13.8	8.9	US shale oil	3.05
Consumption	20,6	18.9	US 2006 - 2014	3.22

Gas (Bcf/day)				
	2006	2013	2014	
Production	50.7	66.5	Qatar	21
Import	8.8	4.8	US shale gas	35
Consumption	59.5	71.3	US 2006 - 2014	15.8

Figure 2 – US oil (top) and gas (bottom) imports and production between 2006 and 2013 (BP 2014 outlook).

How?

Yet US energy dependency alone cannot explain this revolution. The exponential growth of oil and gas prices between 2000 and 2008 in a tense international context where production struggled to satisfy the increasing demand from emerging countries was a second catalyst. In mid-2008, the barrel of oil hovered around the US$ 150 mark[6] while the Henry Hub[7] neared US$ 10/MBTU. Conditions proved highly conducive to launching the massive development of reserves, but the shale oil and gas revolution cannot be explained by these exogenous factors alone. It is also based on four fundamental endogenous 'pillars' (**Figure 3**) typically characteristic of the United States:

- an extremely accurate **knowledge** of the subsurface recognized since the beginning of the XXth century and acquired by virtue of the several million wells drilled. The Americans had therefore already discovered shale oil and gas reserves some time ago, though they

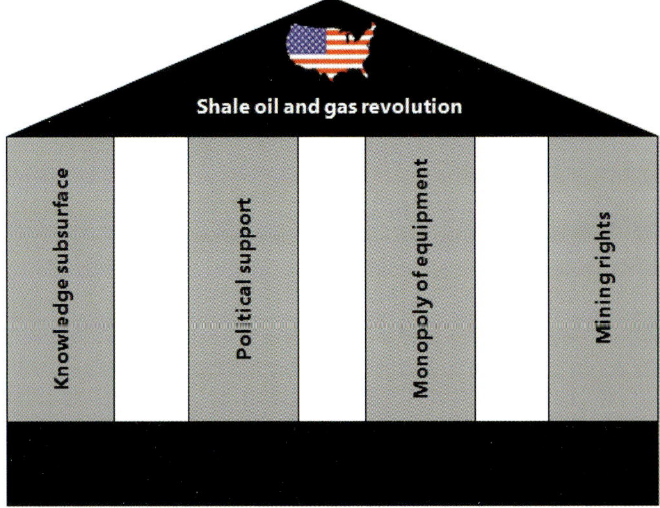

Figure 3 – Knowledge, culture, monopoly of equipment and law are the four main endogenous foundations of the "United States house". They catalyzed the US shale oil and gas revolution.

6. It plummeted in the subsequent months owing to the subprimes crisis.
7. The Henry Hub is the gas spot price in the United States.

were left unexploited owing to their low profitability[8]. By way of comparison, there are only a few thousand wells in Europe;

- a one hundred year-old oil and gas **culture** backed by an unwavering political system. For decades the average American has been used to the sight of drilling rigs and production wells and lives in perfect harmony and continuing trust with the oil and gas operators. The government promotes the consequent opportunities (employment, growth) and accepts the risks. Whatever the administration, American policy has always been underpinned by this culture and has supported the entire activity through incentive-based legislation. In a certain sense, whereas politics in the *"old continent"* tends to restrict economics, in the New World it is politics that adapts to economics;
- a more than favorable mining **legislation** in that the subsurface belongs to the land owner, and not to the State, as is the case in most other countries. For land owners, exploiting the oil or gas is a sizeable financial advantage;
- an almost blanket **monopoly** on world oil and gas equipment. Of the 2,400 drilling rigs in operation worldwide, 80% are in the United States. They also have an abundant service offer in a very open and extremely competitive market. It has a decisive effect on the cost of wells, which represents 70% to 90% of investments.

When it comes to gas, on top of these four essential pillars must be added a very dense pipeline network enabling gas to be distributed throughout the country.

Technically speaking, there is nothing revolutionary about the development of shale oil and gas. It is based on the combination of two mature technologies – horizontal drilling, a technique used on an industrial scale since the beginning of the eighties and hydraulic fracturing, for which the first test dates back to... 1947. The four endogenous pillars mentioned above are the true foundations of the American shale oil and gas revolution. They opened the way for the implementation of *trial & error* developments, highly suitable in terms of both operational (drilling time) and economic (well costs) performance

8. During the 1820s, that is several dozen years before Colonel Drake drilled the first well in Pennsylvania, Fredonia shales were exploited in the state of New York. In the 1920s, a significant part of U.S. gas production came from similar shale deposits in the Appalachian Basin.

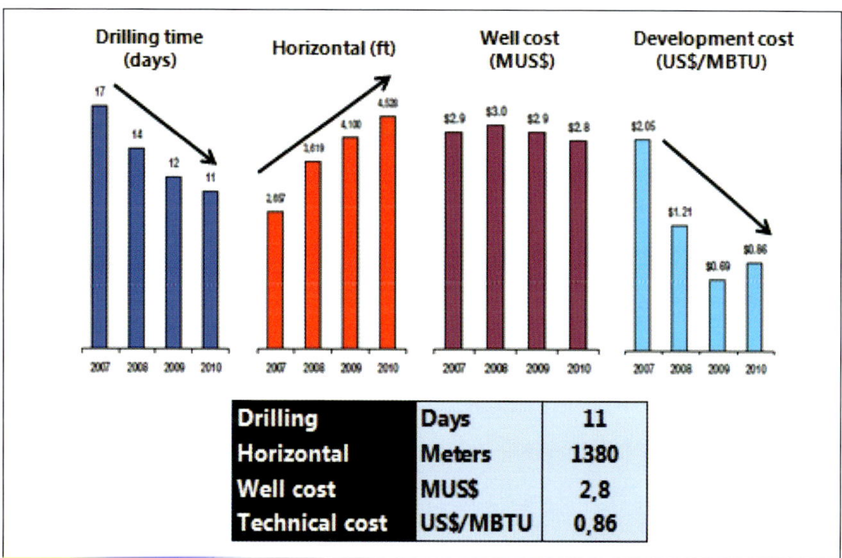

Drilling	Days	11
Horizontal	Meters	1380
Well cost	MUS$	2,8
Technical cost	US$/MBTU	0,86

Figure 4 – The remarkable evolution of operational performance and well costs in the United States[9].

(**Figure 4**). This kind of development consists in drilling and then fracturing a very large number of low-cost wells, without analyzing the geological attributes in detail, and accepting that statistically, a large percentage of them will under-perform. Exporting this model to other countries where the well cost is significantly higher is not really a viable option.

Where?

By the end of 2013, shale oil production represented 27% of US oil production, and shale gas, 42% of its gas production. Yet although the reservoirs are numerous and geographically distributed across the entire continent, the five largest alone represent 65% of oil production and 54% of gas production (**Figure 5**). For oil, the two leaders go by the names of *Bakken* (North Dakota) and *Eagle Ford* (southern Texas) and for gas, *Marcellus* (Pennsylvania), *Barnett* (North Texas) and *Haynesville* (Texas and Louisiana).

9. Data issued from a database of 8,000 wells on approximately 600,000 acres. D. Bentley (Schlumberger) – Doing More with Less - Engaging stakeholders and managing misinformation for shale gas workshop – London 25/09/2012

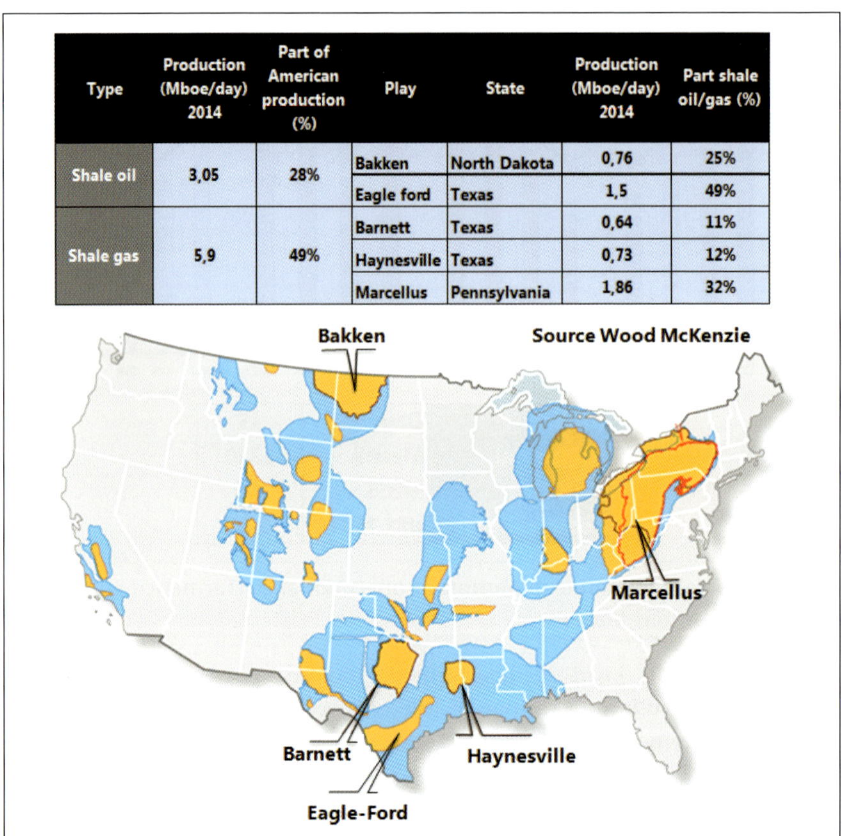

Type	Production (Mboe/day) 2014	Part of American production (%)	Play	State	Production (Mboe/day) 2014	Part shale oil/gas (%)
Shale oil	3,05	28%	Bakken	North Dakota	0,76	25%
			Eagle ford	Texas	1,5	49%
Shale gas	5,9	49%	Barnett	Texas	0,64	11%
			Haynesville	Texas	0,73	12%
			Marcellus	Pennsylvania	1,86	32%

Figure 5 – Distribution of shale oil and gas reservoirs in the United States. The five largest represent 75% of shale oil and 54% of shale gas production.

Where are the world shale oil and gas resources? Which countries will follow the US and with what degree of maturity? What effect will this have on the oil and gas peaks?

All the regions in the world that produce conventional oil and gas have shale oil and gas resources. In its 2013 report, the EIA estimated worldwide recoverable unconventional resources to be 1,200 Gboe for gas and 350 Gboe for oil. Outside North America, the largest gas reservoirs are thought to be in China, Argentina, Algeria and, to a lesser extent, Australia and South Africa, whereas for oil it is Russia that holds the largest resources, followed by China and Argentina. Even though it is not one of the "*top 10*", Europe is thought to have significant gas resources (approximately 80 Gboe) essentially in Poland and France, and to a lesser extent Romania, Great Britain, Denmark and the Netherlands.

In theory therefore, shale oils and gases could double the reserves of conventional gas estimated at 1,100 Gboe and boost conventional oil reserves, estimated at 1,650 Gboe, by 20%. If we factor in these figures, they make a significant difference to world production curves. Peak oil would be pushed back by six years (2026) and the gas peak by almost 20 years (2047).This abundance of new resources should prevent a shortage in the offer in the medium-term. If we consider the median scenario of the International Energy Agency based on a growth in demand of +0.5%/year for oil and of +1.6%/year for gas, unconventional resources would put off a shortage in the supply of oil until 2037 and that of gas to 2045. This scheme would give an energy mix in which fossil energies (oil + gas + coal) would represent 75% of global energy consumption in 2035, i.e. a reduction of 5% compared with the current situation.

These notional evaluations must nonetheless be considered with the greatest caution in that, apart from North America, they are based on simplistic volumetric calculations and do not factor in the economic, political and cultural contexts. Although a thorough knowledge of the geology of a reservoir is the first milestone in the process of extracting reserves, it cannot by itself successfully export the American revolution.

The North American subsurface is far from being unique in terms of the quality of its source rocks. All the regions of the world that produce conventional oil and gas have source rocks, and therefore shale oil and gas resources. From Australia, through Europe and China to Argentina, there are myriad reservoirs, some of which are very promising and of higher quality than those in the United States.

In its latest report published in 2013, the EIA[1] estimates that the worldwide technically recoverable resources are somewhere in the region of 1,200 Gboe for shale gas and 347 Gboe for shale oil. Outside North America, the largest gas reservoirs are thought to be in China, Argentina, Algeria and, to a lesser extent, Australia and South Africa, whereas for oil it is Russia that holds the largest resources, followed by China and Argentina (**Figure 1**). Even though it is not one of the "top 10", Europe is thought to have significant gas resources essentially in Poland and France, and to a lesser extent Romania, Great Britain, Denmark and the Netherlands. These unconventional hydrocarbons could theoretically double conventional gas reserves, estimated at 1,100 Gboe and boost conventional oil reserves, estimated at 1,650 Gbbls[2] by 20%.

These notional evaluations must nonetheless be considered with the greatest caution in that, apart from North America, they are based on simplistic volumetric calculations and do not factor in the economic, political and cultural contexts. Although a thorough knowledge of the geology of a reservoir is the first milestone in the process of extracting reserves, it cannot by itself successfully export the American revolution. Once the geological uncertainties have been lifted, the maturity of a

1. US EIA (Energy Information Administration) "Technically Recoverable Shale Oil & Gas Resources : an assessment of 137 Shale Formations in 41 countries outside the US" June 2013.
2. P.A. Charlez (2014), "Our Energy future is not set in stone", Editions Technip.

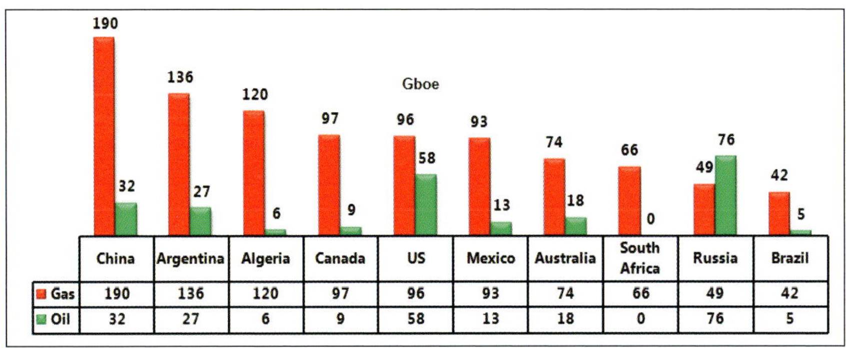

	China	Argentina	Algeria	Canada	US	Mexico	Australia	South Africa	Russia	Brazil
Gas	190	136	120	97	96	93	74	66	49	42
Oil	32	27	6	9	58	13	18	0	76	5

Figure 1 – The top 10 for shale oil and gas resources. Recoverable resources by country (EIA report 2013)

region will depend on a number of different parameters that will either help or hinder future developments. These determining factors include[3]:

- The legal content, in particular mining law[4] and tax regulations.
- The entrepreneurial environment and availability of capital.
- Accessibility to the site and to a source of water, which is sometimes difficult for geographical, demographic and climatic reasons.
- The local presence of operators and oil & gas services (seismic, drilling, fracturing, surface installations).
- The existence of infrastructures (roads) and collection networks for gas.
- Solid political support from regional and local government authorities.
- A positive public opinion where opportunities win out over risks.

A rough division of the *"learning curve"* above into three periods (an initial period to estimate the resources in place, an appraisal period with business pilots being run and an advanced development period - **Figure 2**) plainly demonstrates that outside the United States and Canada, unconventional resources are only in the early stages of development[5].

3. HS CERA (2013), "Exporting the Unconventional Revolution to the World Where, when, and how much?" Global Energy WATCH®.
4. Note that in the US, the legislation is favorable to landowners whereas in most other countries, the subsurface is the property of the state.
5. Outlook for Unconventional Oil & Gas outside North America. Mc Kinsey & Company, June 2013.

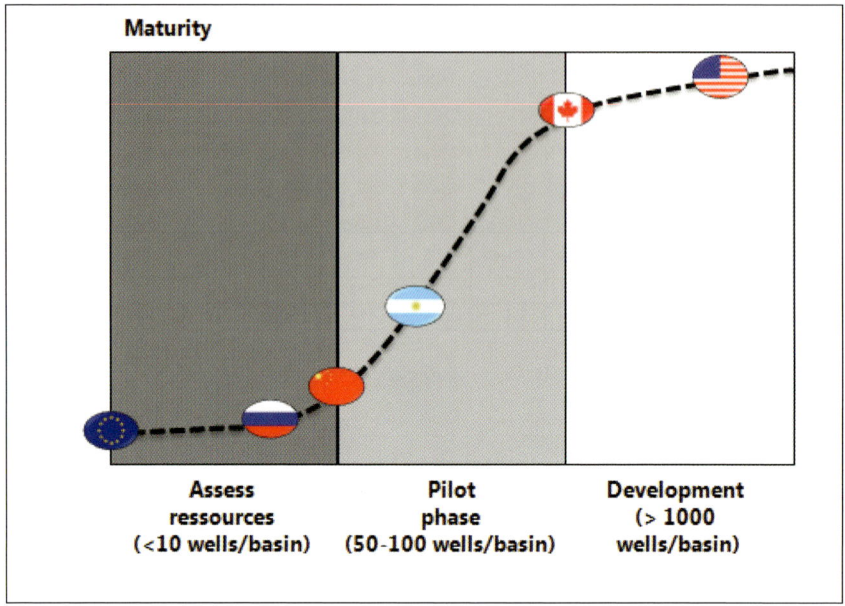

Figure 2 – Learning and development curve for certain regions.

Argentina

Argentina was a gas exporter up to 2004 in a local market that received large subsidies from the State, but in 2011 it imported 120 kboe/day of gas at great expense, essentially from Bolivia and the Middle East (Liquefied Natural Gas). In the medium term, these imports could suffocate the public finances of a country that is still in debt and whose dramatic financial crisis in 2002 still haunts the minds of its population[6]. For Argentina, restricting gas imports has become a strategic objective. The door is therefore wide open to develop the vast unconventional potential that the EIA estimates at 27 Gboe of oil and 136 Gboe of gas. The Neuquen basin (**Figure 3**), which is produced for its conventional hydrocarbons since the 1970s, represents more than half of this unconventional potential. It is *"stored"* in a source rock the quality of which is recognized as one of the best in the world: the *"Vaca Muerta"*.

6. http://www.calpa-paris.org/spip.php?article303

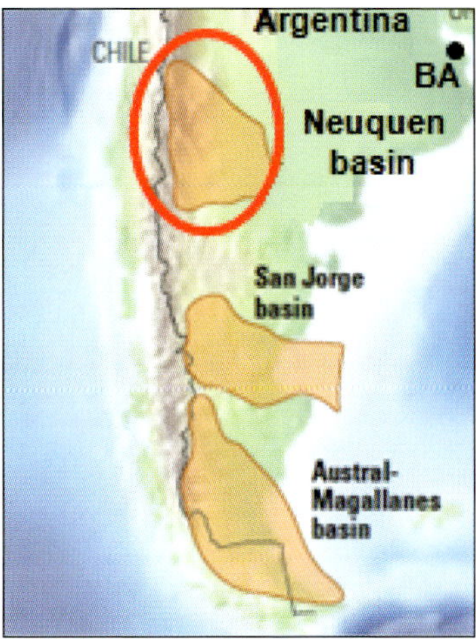

Figure 3 – The Neuquen basin (Argentina) and its famous *Vaca Muerta.*

At its widest, the thickness of the source rock reaches 600 m. The existence of infrastructures constructed to export conventional resources is a major economic advantage for the future development of the source rock, impregnated with both oil and gas. Despite its attractive geology, however, the economic conditions (strictly regulated gas prices) and the political climate (nationalization of YPF in April 2012[7]) could be potential stumbling blocks. Currently in the pilot phase, the *Vaca Muerta* is the most advanced unconventional project outside North America. It is thought that its development would require the drilling of several tens of thousands of wells and the mobilization of one to two hundred drilling rigs.

China

With 200 Gboe of shale gas and 32 Gboe of shale oil, China is considered as having an unconventional potential comparable to that of North America. The resources are located chiefly in the Tarim basin in the North West and the Sichuan basin in the South East (**Figure 4**).

7. http://www.rfi.fr/economie/20120417-argentine-expropriation-ypf-derniere-une-longue-serie-nationalisations

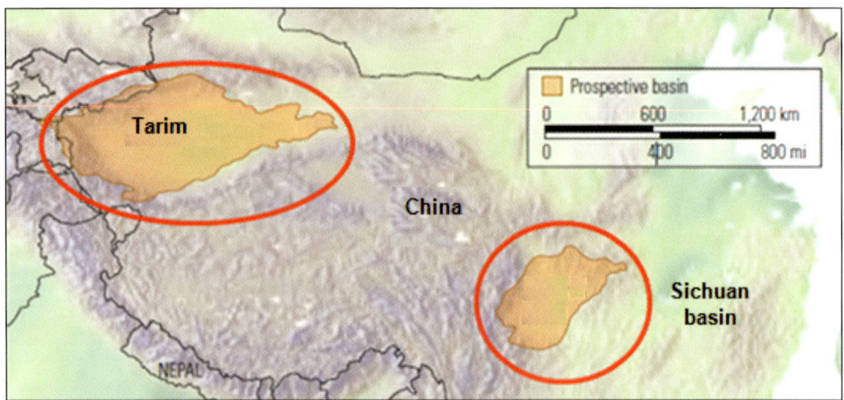

Figure 4 – The Sichuan and Tarim basins the two giant Chinese reservoirs.

Although the initial wells drilled have confirmed the potential of these basins, rapid commercialization of the resources is not an easy task. Given the considerable average depth (between 3,000 m and 4,500 m), associated with highly-fractured geological structures and significant uncertainties as to future production, it would be difficult to make these resources competitive for anything less than US$ 12/MBTU[8]. Moreover, in spite of the fact that it is recognized as one of the largest basins in the world by its sheer size (400,000 km^2), the Tarim is located in an arid region (much of the basin is covered by the Taklamakan desert) bordered by very high mountains (Tian Shan to the North, the Pamirs chain to the West and the Kunlun to the South) where the water supply required for hydraulic fracturing, would be one of the main obstacles. The production of unconventional gas should enable China to restrict its LNG imports and also significantly decrease its consumption of coal, which is still by far its dominant primary source of electricity generation (12.8 Gboe of coal per year compared with 3.65 Gboe of oil and less than one Gboe of gas in 2012). China is responsible for more than a quarter of the global greenhouse gas emissions.

8. IHS CERA (2013), "Exporting the Unconventional Revolution to the World Where, when, and how much?" Global Energy WATCH®.

Russia

Russia's unconventional potential lies essentially in Bazhenov, the source rock for the conventional reservoirs of Western Siberia which in 2012 were producing 7.5 Mboe/day, i.e. 75 % of Russian production. Stretching out over a gigantic surface of almost a million km², Bazhenov is the largest petroleum basin in the world. Located at a depth of between 2,550 m and 3,000 m, it is estimated to hold reserves of 75 Gboe of oil and about 50 Gboe of gas according to the EIA. Its southern part is impregnated with excellent quality oil whereas the northern sector is impregnated with gas and condensates. In spite of the fact that 2,500 wells have been drilled there since the 1960s, and that up to now the basin has been considered as uneconomic, it is still underdeveloped and was producing just 13.5 kboe/day in 2012. The geology is heterogeneous and complex, and there is a high risk of drilling poor production wells. Although it is located in a region with a hostile polar climate (in winter, temperatures fall to below -40°C, and the permafrost makes operations particularly difficult and costly in spring and summer), Bazenhov has the great advantage of benefiting from existing infrastructures. Numerous local operators, service companies and research centers set up in the region some time ago. The Russian authorities, who wish to compensate for the decline of conventional fields, have begun a legislative process designed to offer attractive tax conditions. The uncertainties can be lifted only once the pilot phases have been launched and a sufficient number of wells has been drilled.

Europe

According to the latest EIA evaluations (**Figure 5**), Europe is thought to have 93 Gboe of unconventional resources, of which 85% are gas resources. The main ones are located in Poland and France (each having 28 Gboe) and the others in Romania, Denmark, the Netherlands and Great Britain. The other prospects (Germany and Sweden) offer only marginal resources.

However, these estimates based on simplistic volumetric calculations, are still inaccurate and must therefore be considered with a great deal of caution. Only the data acquired from exploration wells will give companies

Country	Gas Gboe	Oil Gboe	Total Gboe
Poland	25	3,3	28
France	23	4,7	28
Romania	9	0,3	9
Denmark	5	0	5
United Kingdom	4	0,7	5
Holland	4	2,9	7
Others	9	1	10
Total	80	12,9	93

Figure 5 – Estimated shale oil and gas resources in Europe (source: US Energy Information Administration).

an idea of the actual stakes involved.[9] If we assume that these hypotheses were to be confirmed at least in part, the exploitation of shale gas in Europe, though not a game changer, could reduce the continent's energy dependency on its main gas suppliers (Norway, Russia and Africa) and have significant consequences on growth, energy prices and employment.

Impact on peak oil and peak gas

As mentioned previously, conventional reserves today are estimated at 1,560 Gboe for oil and 1,100 Gboe for gas which, in 2005, led experts to consider peak oil at a little over 90 Mboe/day by 2020 and peak gas at around 70 Mboe/day five years later (**Figure 6**).

The consideration of unconventional resources (350 Gboe for oil and over 1,000 Gboe for gas) would significantly impact production curves. The oil peak would therefore be pushed back by six years (2026) and raised to 107 Mboe/day, whereas the gas peak would be delayed for almost 20 years (2047) and ramped up to 100 Mboe/day. This abundance of new resources (between 6 and 7 Mboe/day of oil reserve capacity by 2015[10]) should put off a shortage in supply. If we consider the alternative scenario

9. Thomas Moore Institute (2012), "Gaz de schiste et Europe. Analyse comparative dans 14 pays Européens", Benchmarking Memo.

10. BP Energy Outlook 2030.

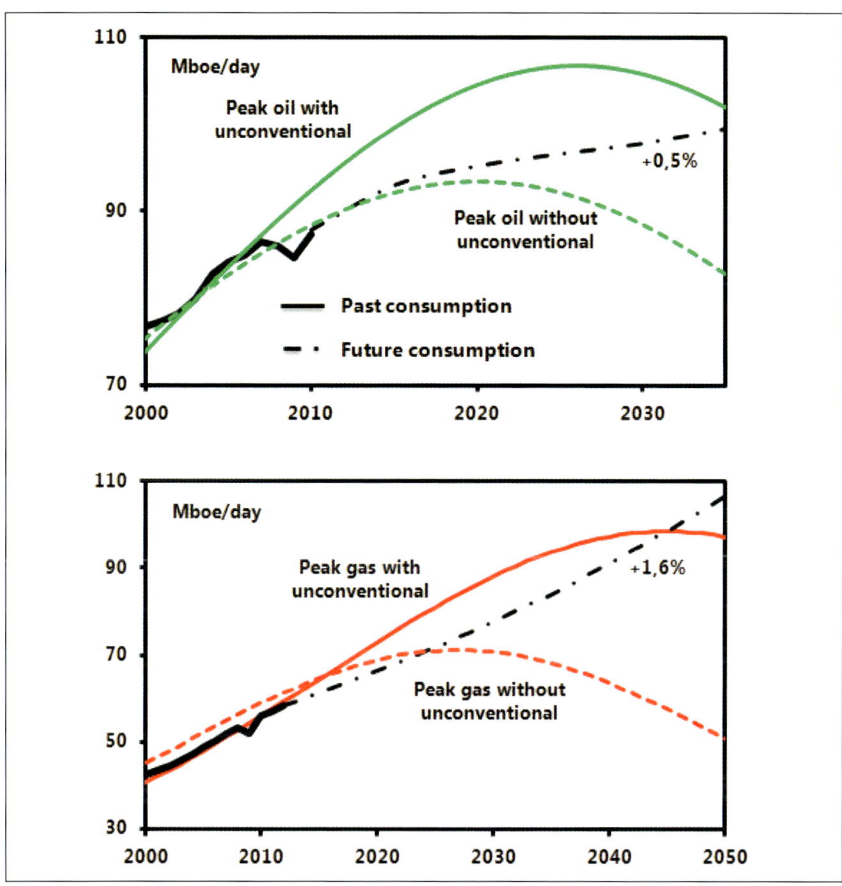

Figure 6 – Revised peak oil and gas (top and bottom respectively) based on the King Hubbert[11] model, factoring in shale oil and gas resources from the EIA and the demand curves proposed by the IEA.

from the IEA[12] based on a growth of +0.5% for oil and +1.6% for gas, that takes us to an oil demand of 100 Mboe/day and a gas demand of 84 Mboe/day by 2035. In this case, unconventional resources would postpone the oil supply shortfall to 2037 and that of gas to 2045. This scheme would give an energy mix in which fossil energies (oil + gas + coal) would

11. See technical appendices BP Energy Outlook 2030.

12 International Energy Agency - IEA outlook 2013.

represent 75% of global energy consumption in 2035, i.e. a reduction of 5% compared with the current situation (**Figure 7**).

	Gboe	Part	
Oil	134	27%	76%
Gas		24%	
Coal		25%	
Wood		11%	24%
Nuclear		7%	
Hydro		3%	
Renewable		3%	

Figure 7 – Energy mix in 2035 (source: EIA outlook 2012)

What consequences will the American revolution have on the world economy in the medium term?

The United States seized the opportunity of the shale oil and gas revolution to become one of the most competitive countries in the world. Its recovery was due first and foremost to the crash in gas prices which, in mid-2012, dropped below US$ 2/MBTU. At the same time the World Trade Index broke away from the Brent Index, a true godsend for American consumers.

These decreases did not help oil and gas operators, but it did boost activities related to refining, petrochemicals, chemicals and the "*energy-guzzling*" sectors like the steel cement and glass industries as well as those involved in gas transport and electricity generation. The US petrochemical industry halved its operating costs while on the other side of the Atlantic, Europe doubled them. In the medium term, fears that the European refining and petrochemicals industries would relocate to the US are well-founded, as between 2006 and 2012, 1.75 million jobs were created there.

The decrease in gas prices also forced the United States to switch their coal-fired electricity generation to gas, a move which reduced their greenhouse gas emissions by 13%.

Over the next fifteen years, the US should continue to ramp up their production of shale oil and gas. The growth of the petrochemical and manufacturing industries should generate 2.3 million jobs in 2015 and 3.5 million in 2035. The transition in electricity generation should also follow its course leading to a further reduction in GHG emissions.

Even if they are in a position to be gas exporters as from 2020, the US will never be self-sufficient in terms of oil.

This unexpected panorama will have major geopolitical consequences. In 2035, US oil imports from the Gulf will represent no more than 15% (compared with 64% in 2000) whereas China and India, which were just 14% in 2000, will reach 75%.

The second upheaval concerns the global reorganization of Liquefied Natural Gas (LNG) flows. In the early years of this century, the US was set to become a massive LNG importer, but this situation has completely changed. In the space of 20 years, the shale gas boom could enable the United States to produce a quarter of global LNG production. For Russia and its gas giant Gazprom, this new state of affairs represents a threat, as US LNG is competing with the Russian gas exports destined for the European Union. It is therefore to China and India, which are planning to move their source of electricity generation from coal to gas, that Russia will turn.

Finally, the upheaval in gas flows has already had major consequences on those for coal. In the United States, the electricity generation switchover caused coal prices to plummet, encouraging certain European countries, and Germany in particular, to reopen coal-fired power stations to satisfy their political commitment to abandon nuclear energy by 2022. This paradoxical strategy threatens to jeopardize the Union's environmental targets regarding greenhouse gases.

The United States seized the opportunity presented by the shale oil and gas revolution to come out of the subprime crisis with its head held high. In September 2008, it had caused global chaos and, less than four years later, it had once again become one of the most competitive countries in the world.

Gas prices collapse and oil prices break away

The renewed competitive edge is due first and foremost to the collapse of gas prices on a market taken completely by surprise by this sudden influx. Perfectly in step until early 2009, the world oil prices and those of US gas[1] began to diverge considerably. By mid-2012, when the price of oil remained above US$ 100, that of American gas fell to below US$ 2/MBTU[2].

1. The famous Henry Hub.
2. MBTU = Million British Thermal Units. 1 MBTU is approximately equivalent to 0.17 boe. At the beginning of 2012, gas prices in Japan were almost the same as those for oil (US$ 100/boe).

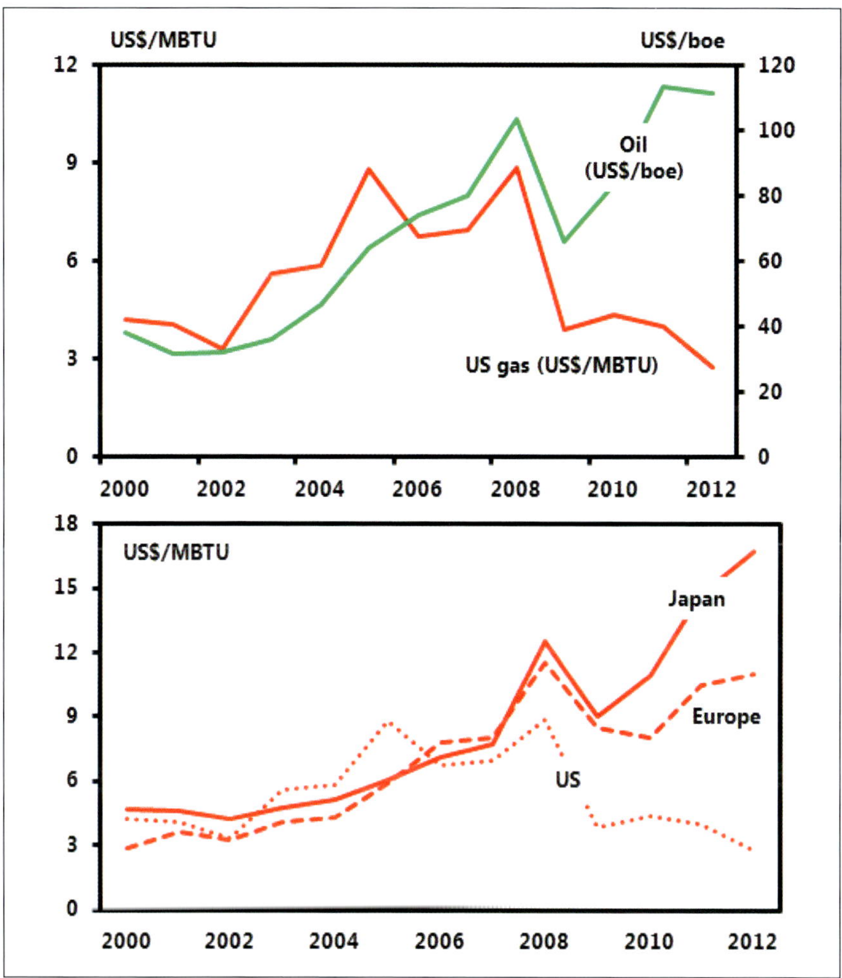

Figure 1 – Spot gas prices from 1999 US/Europe/Japan

Regarding oil, remember that if unconventional sources had not been developed, global capacity would now be limited and, when the market began to recover in 2009/2010, would certainly have caused prices to soar. For the time being, more than anything else, shale oils have helped smooth prices out. Looked at in perspective, they have also contributed to regionalizing the markets – a first in the history of oil. That meant that between the end of 2010 and June 2013, oil prices in the US (the WTI) broke away from those on the Brent index, with differences of more than $20/bbl, a godsend for US consumers (**Figure 2**).

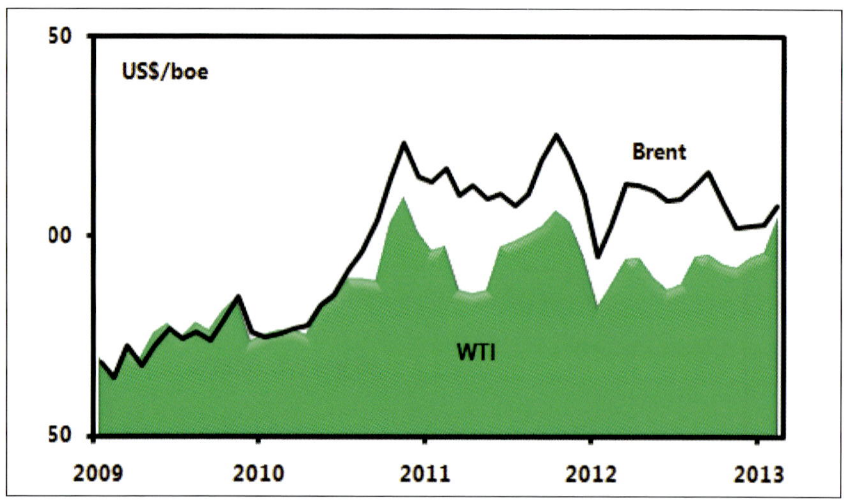

Figure 2 – Evolution of the Brent index and the WTI (source EIA[3])

In the longer term, discussions about prices cannot be based solely on a debate about unconventional hydrocarbons, but must be considered from a global perspective. The cost of developing shale oils (US$ 60 to US$ 80/bbl) does not allow for a significant drop on the stock exchange. Such an event, if it did not stop the production of shale oils completely, would, at best, almost immediately and significantly slow it down.

Those who win out in the short term… are not the ones you might expect

Contrary to what would might have been be expected, it was not the Upstream oil sector (Exploration & Production), to a certain extent *"victim of its own success"* which reaped the benefits from the marked decrease in gas prices, but rather activities related to refining, petrochemicals and chemistry.

But the out-and-out winner of the shale oil and gas revolution is the Midstream sector (gas transport and electricity generation) which, considerably in deficit between 2000 and 2008, has seen margins explode over the last three years (**Figure 3**). Other "energy-guzzling" manufacturing

3. http://www.eia.gov/dnav/pet/hist/LeafHandler.ashx?n=pet&s=rbrte&f=m

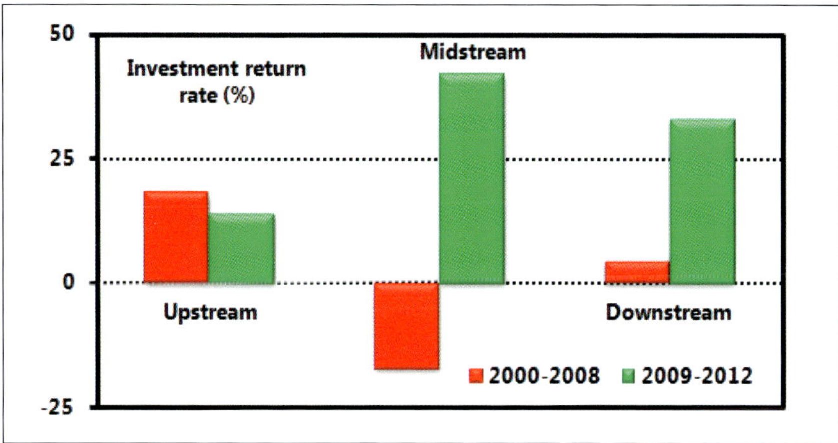

Figure 3 – Economic impact on the up/mid/down stream (source: McKinsey Corporate Performance Analysis Tool).

sectors like steel, cement and glass also reaped considerable benefits from the decrease in stock prices, significantly improving their margins. This spectacular revival triggered a domino effect in the industrial sector as a whole, with numerous jobs being created. According to IHS Global Insight[4], over the period 2006 to 2012 1.75 million jobs were created in the US. Above and beyond the activities directly linked to oil and gas production, the decrease in energy costs put a stop to outsourcing and generated employment in many sectors. In addition to refining and the chemical industry, jobs have also been created in mining, civil engineering, transport and manufacturing. The jobs created are usually very high-quality, well paid and have positive consequences in terms of consumption. In 2010, the contribution of unconventional resources to the GDP in the US was estimated at US$ 77 billion.

Serious consequences for chemicals and refining, particularly in Europe

As gas represents over 50% of operating costs, gains in competitiveness in the US petrochemical industry (which is the basis for plastics manufacturing) over the period 2006-2012 were spectacular (**Figure 4**).

4. "The economic and employment contributions of shale gas in the United States," IHS Global Insight, Dec 2011.

Whereas in 2006 production costs in US/Europe were very similar[5] (0.5 to US$ 0.6/pound), in 2012 US production costs had been halved, while over the same period, production costs in Europe had almost doubled. European production had become less attractive and stagnated, while across the Atlantic, the US were becoming an unprecedented *"land of growth"*[6] quadrupling their production capacity over the same six years. If this production cost difference persists, it is possible that the European refining and petrochemical industries, whose margins are currently negative, could relocate to the US, causing extensive job losses in Europe.

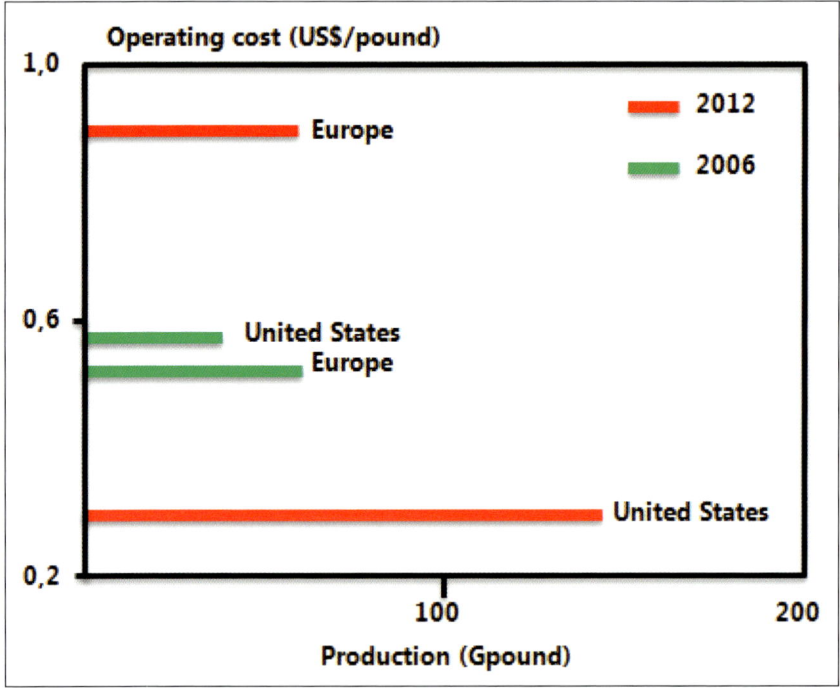

Figure 4 – Evolution of the competitiveness of the petrochemical industry between 2006 and 2012 in the US, and Europe (Source IFRI).

5. American costs were even slightly higher than European costs.
6. C. Maisonneuve (2013), "Hydrocarbures non conventionnels : quelles conséquences économiques?", BRGM Science Days, October 13, 2013.

Beneficial effects on greenhouse gas emissions

The decrease in gas prices compelled the United States to change their electricity generating source from coal to gas. A significant number of coal-fired power stations were therefore shut down, consequently causing prices to crash. The ton of coal, which had reached a peak price at US$ 120/ton in 2009, was being negotiated at US$ 65/ton a year later and at the same time, the gas-fired power stations were reopened. The displacement had an extremely positive impact on greenhouse gas emissions which, having fallen from 6 Gtons in 2004 to 5.25 Gtons in 2012, were reduced by 13% (**Figure 5**).

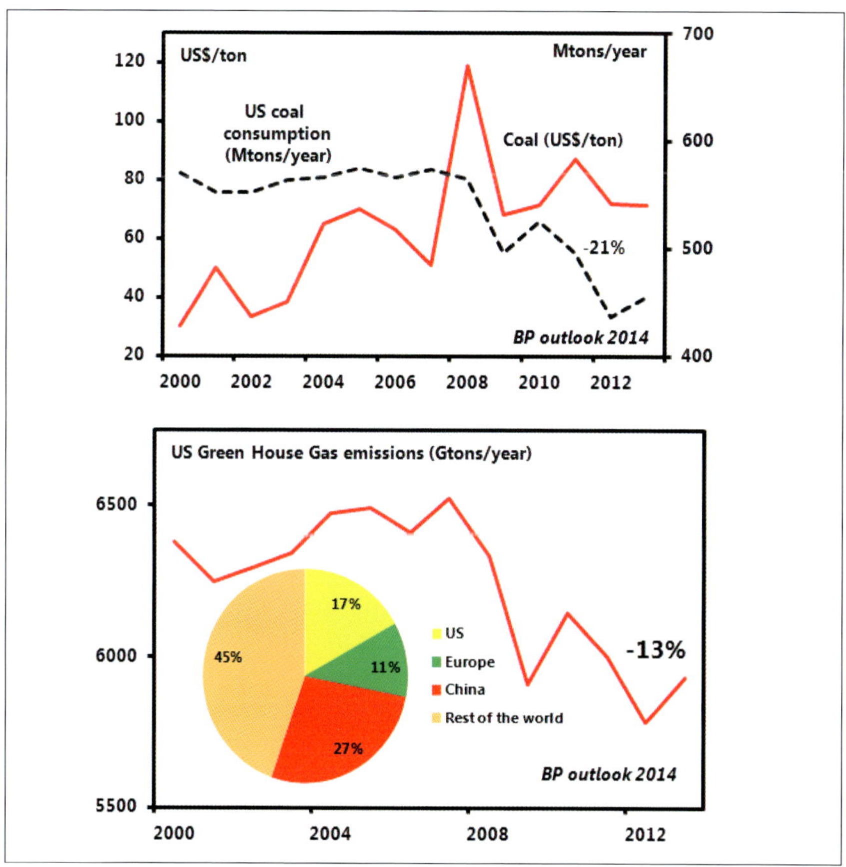

Figure 5 – Collapse of US coal prices. Impact of the displacement from coal to gas on US emissions (source: EIA)

In the near future, without having signed the protocol, the United States could fulfil their Kyoto objective while emissions levels in Europe, which is remobilizing its coal-fired power stations to compensate for its commitment to stopping nuclear developments (in particular in Germany) in the short term, are starting to climb.

Vigorous and optimistic medium-term growth

Although they are the undisputed economic winners, at least in the short term, of the energy revolution that they instigated, are the US able to continue this unprecedented growth and reach energy self-sufficiency that some announce for 2020 - 2025[7]?

According to the IEA, the considerable potential of shale oil and gases should enable the US to significantly increase their hydrocarbon production over the next fifteen years (**Figure 6**). By 2035 the production of shale gases could reach 7.3 Mboe/day (i.e. 50% of American production). As for oil production, forecasts predict a second flat oil peak at 11 Mboe/day[8] between 2020 and 2025. However, contrary to the rumors promulgated by certain media[9], as things stand today, if the US were to become gas exporters as from 2020, oil self-sufficiency does not seem feasible. Coupled with a significant decrease in consumption (between 0.5% and -1.5%/year), in the best case scenario, the production of shale oils would help the US reduce their oil dependency to 25%, imports falling from 11.4 Mboe/day in 2010 to just 5.4 Mboe/day in 2020 and to 4 Mboe/day in 2025.

If all the predictions are confirmed, unconventional hydrocarbons would contribute respectively US$ 120 billion in 2015 and US$ 230 billion in 2035[10] to American GDP, generating along the way respectively 2.3 million jobs in 2015 and 3.5 million jobs in 2035[11]. Stimulated by attractive gas prices, growth in the downstream sector and in particular

7. N. Baverez (2013), "La révolution énergétique américaine", Le Figaro Chronique, June 26, 2013.
8. i.e. almost equivalent to that in 1970.
9. http://oilprice.com/Finance/investing-and-trading-reports/USA-Energy-Independence-Sense-or-Nonsense.html
10. "The economic and employment contributions of shale gas in the United States", ISH Global Insight, Dec 2011.
11. IHS global insight.

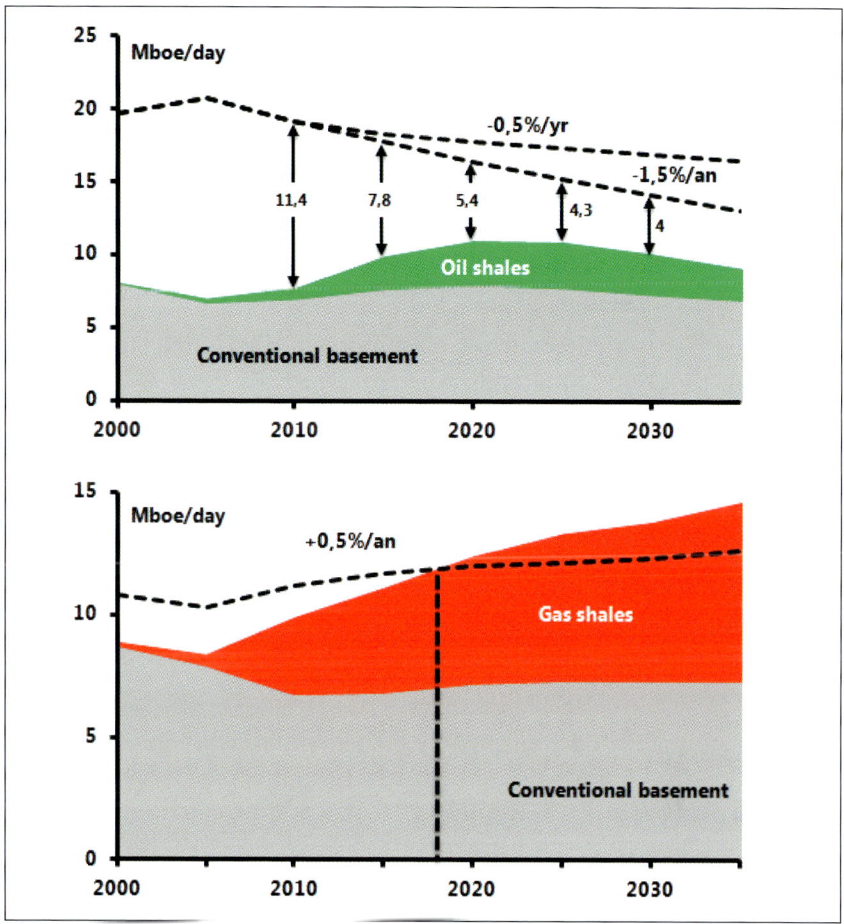

Figure 6 – Production history and forecasts in the US.
Sources: BP 2013 outlook, IHS CERA 2012

petrochemicals should continue. The ethylene, methanol and ammonia channel should therefore also significantly increase its capacity by 2020[12], just like the manufacturing industry, which could increase its production by 2.9% in 2017 and 4.7% by 2035[13].

12. IHS CERA - Robert Ineson - Personnel communication – Houston January 29, 2013.
13. "Shale gas: A renaissance in US manufacturing ?", PWC, December 2011.

The shift in American electricity generation from coal to gas should also continue on a massive scale. By 2020, 44GW would have been displaced, which corresponds to an extra gas demand of 0.5 Mboe/day.

Major consequences for oil, gas and coal

The mass production of shale oil and gas in the United States should, in the medium term, significantly alter hydrocarbon flows, and these changes in turn will affect the geopolitical equilibria worldwide.

Between 2010 and 2030, while oil imports in the United States would be halved, those in China and India will triple. As these mass imports come essentially from the Middle East, China and India would in fact become the main clients of the Gulf monarchies (**Figure 7**). The figures are startling: in 2035, US oil imports from the Gulf will represent just 15% (compared with 64% in 2000) whereas those of China and India, which were just 14% in the year 2000, will top 75%. Even if the scenario of a abrupt withdrawal by the Middle East does not seem credible, the United States, partly delivered from their energy dependency on the Gulf oil monarchies, should gradually lose their influence, to the benefit of China and India.

The second shake-up concerns the global reinterpretation of Liquefied Natural Gas (LNG) flows, for which world production in 2012 was 250 Mtons/year[14]. Qatar (75.5 Mtons – 31% of world production), Malaysia (25 Mtons), Indonesia and Australia (20 Mtons each) were the main producers whereas Japan (78.8 Mtons), South Korea (35.8 Mtons), the UK and Spain (a little less than 20 Mtons each) were the main consumers. In 2005, the EIA was predicting the inevitable decline of American gas production and forecast that to offset this phenomenon, the US would need to import LNG in vast quantities. In 2010 with 23%, the US would become the second client of the world market after Japan. To achieve this objective, the US multiplied their regasification capacities by a factor of seven between 2002 and 2010. Within this scenario, Russia, Iran and Venezuela, all historical adversaries of the United States, were to play an increasingly important role in LNG supply, with a strengthened negotiating position. Mass LNG imports in the US would have had a

14. Source: US DOE.

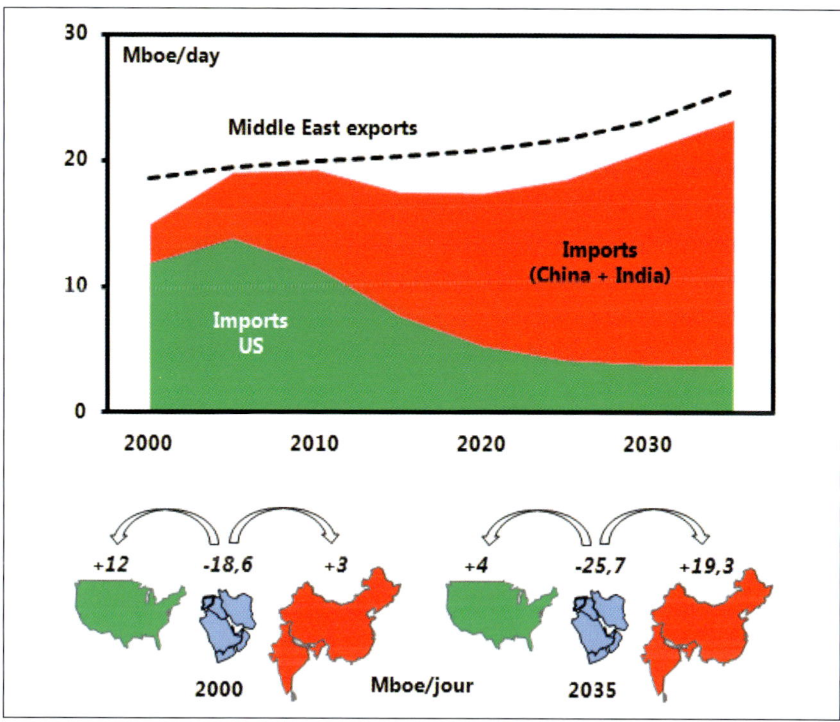

Figure 7 – Evolution of oil flow between the Middle East, China/India and the United States.
(Sources: BP outlook 2013 and IEA outlook 2012)

major impact on prices, making coal more competitive with negative consequences in terms of greenhouse gas emissions. In fact, what happened was the complete opposite.

In 2011 American LNG imports, which reached 5.9 Mtons/year, represented just 2% of the world market. This reversal is set to become more acute over the following years and in particular in 2017 when the US will become net gas exporters. The startup of several liquefaction projects has been announced for 2017 and 2018. This entire state of affairs could boost the United States to reach over 40 Mtons of LNG by 2025 and approach 100 Mtons ten years later, i.e. 25% of world LNG production (**Figure 8**). If for the time being the relentless inversion of American demand has been assured by a post-Fukushima Japan, could the *"wave"* cause overproduction, a significant drop in prices and make the

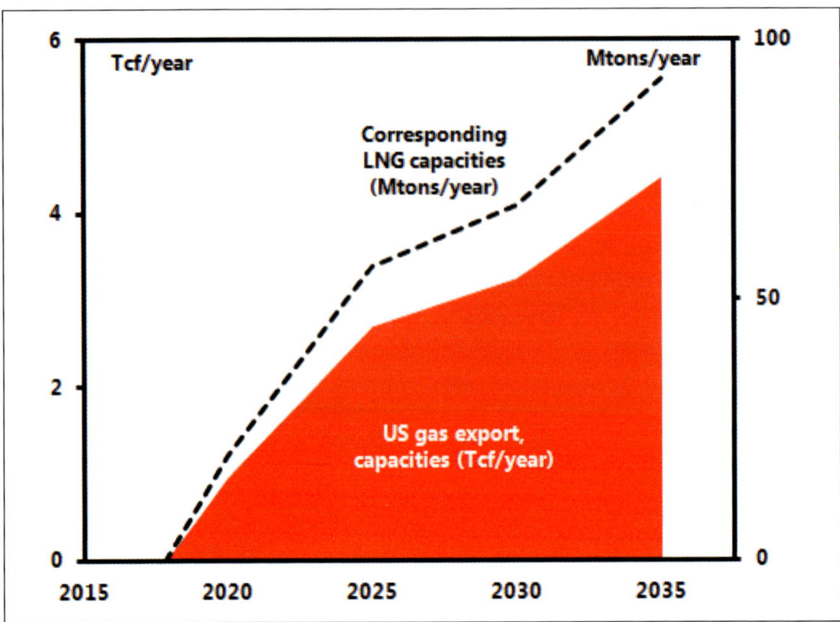

Figure 8 – Potential gas export ramp up in the US Corresponding LNG capacities (Source: IEA)

development of certain highly capitalistic projects in the Arctic[15] or in the deep offshore nonviable? From many perspectives, this new layout makes the Russian strategy, centered essentially on its gas-giant, Gazprom, which alone controls 70% of Russian gas reserves, i.e. 13% of world reserves, particularly vulnerable[16]. In addition to its extensive capitalization, higher than that of ExxonMobil, Gazprom is a powerful lever of economic and political influence that the Russian State has used on several occasions as a means of pressure on its neighboring countries. As 75% of Russian gas exports are now destined for the European Union, it is easy to understand that the potential export of US shale gas at very competitive prices to markets that were traditionally the prerogative of Russia, represents a threat for Gazprom and therefore, indirectly, for the Russian State. So it

15. IGU 2011 World LNG Report. http://www.igu.org/sites/default/files/node-page-field_file/LNG%20Report%202011.pdf

16. Russian reserves, estimated at 1,162 TCF represent approximately 18% of world gas reserves. BP 2013 outlook.

is essentially to China[17] and to a lesser extent, India that Russia would have to look and gradually develop a Sino-Russian axis. However, China could also decide to develop its own shale gas resources[18] , estimated by the EIA at somewhere around 200 Gboe.

Finally, as we have already seen on several occasions, the shake-up of gas flows would have major consequences on those of coal. The shift in electricity generation in the United States triggered a marked drop in coal export prices, encouraging certain countries, in particular Germany[19] to reopen coal-fired power stations to uphold their commitment to completely abandon nuclear energy by 2022. Germany, true to European tradition, is going its own way and is gradually implementing a paradoxical strategy that threatens to compromise the Union's environmental objectives regarding greenhouse gases (**Figure 9**).

Oil flow from the Middle East to India and China; Chinese companies replacing the Americans in the Middle East; gas and coal flows from the US to Europe; the United States meeting emissions targets that it has not signed and Europe falling short of them in spite of itself; Russia sanctioned[20] and at a loss for external outlets for its gas in a flooded LNG market: our energy future is not set in stone[21].

---

17. The gas demand has doubled since 2007, rising from 7 to 14 bcf/day. Gas however, represents just 4.5% of the Chinese domestic gas demand.

18. See question 2.

19. http://www.strategie.gouv.fr/content/la-transition-energetique-allemande-est-elle-soutenable-note-danalyse-281-septembre-2012

20. t.newtondunn@the-sun.co.uk

21. P.A. Charlez (2014), *Our energy future is not set in stone*, Editions Technip, Paris.

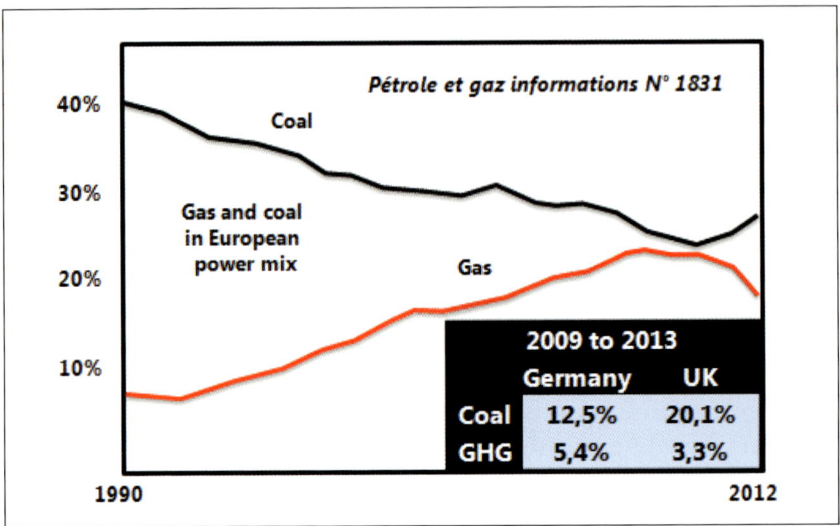

Figure 9 – The European energy paradigm.
UK and Germany have increased their coal consumption by closing gas power plants and reopening coal power plants[22]

22. G. Hureau (2014), « La concurrence gaz/charbon dans le secteur électrique Européen », *Pétrole et Gaz information*, n° 1831 Juillet – Août 2014.

How does nature go about making oil and gas?

Oil and natural gas belong to the family of hydrocarbons, chemical compounds that are a careful blend of hydrogen and carbon. Though the latter are the base elements of many minerals, they are also some of the main ingredients of life. Whether animal or vegetable, "*organic matter*" is rich in carbon and hydrogen and it is this raw material that nature uses to manufacture oil and gas.

The manufacturing process begins with the very gradual accumulation (a few hundredths to a few tenths of a mm per year) of organic matter on the ocean floor. It mixes with a mineral mud made of clay, sand and limestone to make an organic mud. Over time and by deposition, the "*organic mud*" start to bury, hardens and is progressively transformed into rock. With burial, temperature increases on average by one degree every 30 meters.

At medium burial depth, organic matter undergoes an initial transformation, akin to biological fermentation. Bacteria feed on some of the buried organic matter and, once they have digested it, release natural gas, called "**biogenic gas**". The solid residue, called kerogen, is the basic raw material for oil and gas.

For increasing burial depths and temperatures of between 80°C and 150°C, kerogen undergoes thermal cracking which results first of all in oil that is more and more fluid. Over 150°C, organic matter continues to undergo cracking to become natural gas, the scientific name for which is "methane". These hydrocarbons formed by thermal cracking are called "**thermogenic oil and gas**".

The final solid residue which is very rich in carbon is simply coal dispersed in the rock in the form of solid micro-clusters that cannot be produced. But real coal seams can only be formed from a rock whose organic matter content is much higher than its mineral content. To manufacture coal, large quantities of vegetable debris need to accumulate within a shallow layer of water called a peat bog. As the temperature increases, a burial and baking process very similar to that described above for oil and gas, leads to the formation of compounds that are increasingly rich in carbon: peat (50 to 55 %), lignite (55 to 75 %), coal (75 to 90 %) and finally, anthracite (> 90 %). But the hydrogen present in the organic matter also generates gas called "coal bed methane".

Oil and natural gas both belong to the family of hydrocarbons, complex chemical compounds formed from a mixture of hydrogen (hydro) and carbon (carbide).

Hydrogen is the lightest atom in the periodic table of the elements[1]. It is the main constituent of the Sun and most of the stars. Associated with oxygen, it is abundant at the Earth's surface in the form of water (the chemical formula of water is H_20). But as it is too light and the pull of the Earth's gravity is not sufficient to retain it, it escapes naturally from the atmosphere[2] and only exists on the Earth in the form of atomic traces. And it's a real shame, as hydrogen would be the most ecological fuel because, when it is burned in the presence of oxygen, all it produces is... water[3]. Hence the name hydrogen, which means "*water-producing*", in Greek.

Carbon, however, is present in its natural form in the Earth's crust in different minerals. At ambient temperature and pressure, it is black and crumbly: graphite. Used for centuries for writing under the incorrect name of "*pencil lead*", graphite is also used in electrical construction, to make brushes for motors, as an adsorbent for cooker hood filters and even in medicine to treat food poisoning. At depths of about 200 km, when it is under very high pressure, (5,000 times atmospheric pressure) and very high temperature (over 1,000°C) graphite crystallizes and becomes diamond, the hardest known substance in the Earth's crust.

Carbon and hydrogen are the base elements of many minerals, but they are also the main ingredients of life. By associating carbon, hydrogen, oxygen and nitrogen atoms in very long helicoidal chains (the famous

1. Also called the Mendeleev table. See Appendix 2.
2. It was used in airships like the *zeppelin* before being replaced by helium which is less dangerous as it is not flammable.
3. $H_2+1/2\ O_2=H_20$

Figure 1 – Carbon in mineral form (graphite, diamond)

DNA[4]) nature created the first living organisms. And for about a billion years now[5] the ingredients have barely changed. Whether animal or vegetable, any substance (tree wood, plant stems and leaves, seaweed, soft-bodied mollusks, fish bones or the muscles and bones of mammals) made by living beings[6] contains, in addition to water, matter that is particularly rich in carbon and hydrogen – *biomass*. And it is from this raw material, also called *organic matter*, that nature, following an extremely precise *recipe* and with care and determination, manufactures gas and oil. Oceans produce living matter at their surface (where light can penetrate) in abundance, and in particular near Arctic continents and seas[7].

The manufacturing process (**Figure 2**) begins with the accumulation of organic matter from dead animals, but especially vegetable and micro-vegetable debris (microscopic algae also called phytoplankton)[8] which are deposited very slowly on the ocean floor - at the rate of a few hundredths or tenths of mm per year. It mixes with a mineral mud composed of very fine grains of clay, sand and limestone. Together they form an organic mud that, in proportion, contains just a few percent of organic matter submerged in 90% mineral mud. Over time, and with the deposition of underlying sediments, the organic mud starts to be buried. It expels water,

4. Deoxyribose Nucleic Acid
5. The first living organisms are thought to date from the pre-Cambrian period.
6. 90 % of a vegetable, 65 % of a human being.
7. Ph. Ungerer & B. Durand (1987), "La géochimie organique pétrolière", Bulletin de la Société Géologique de France, n°7.
8. Mainly vegetable debris. Debris from vegetal origin only represent one per cent.

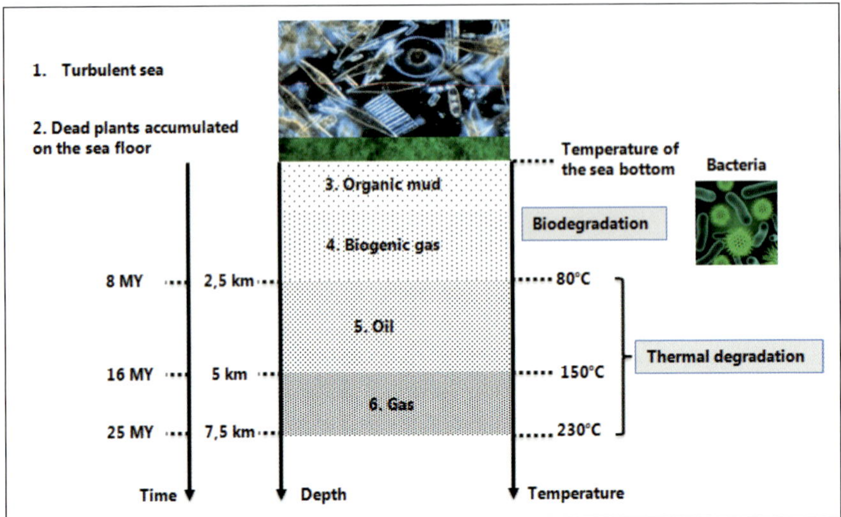

Figure 2 – Oil and gas manufacturing process (the assumed geothermal gradient is 30°/km and the deposition rate is 0.3 mm/year)

hardens[9] and gradually turns into rock. The burial is also accompanied by a proportional increase in temperature[10] of about thirty degrees per km. To simplify things, if we assume a regular deposition rate of three tenths of a millimeter per year, after 1,000 years the biomass will have been buried at a depth of just 0.3m. To reach a depth of 30m and see its temperature rise by one degree would take one hundred thousand years. When the organic mud reaches a depth of 1 km, over three million years will have passed and its temperature will have increased by just 30°C.

At a low burial depth, organic matter will undergo an initial transformation of biological fermentation. Micro-organisms (bacteria) that do not need oxygen[11] and which can survive temperatures of approximately 80°C[12] will feed on part of the buried organic matter and,

9. This hardening phenomenon which, by expelling water, converts mud into water, is called diagensis in geology.

10. The increase in temperature with depth is called the geothermal gradient. It lies somewhere between 15°C/km and 50 °C/km, and on average, is something like 30°/km.

11. This is why they are called "anaerobic" which means "without air" in Greek.

12. 80°C is, as if by chance, the pasteurization temperature above which all living bacteria in milk are destroyed.

Figure 3 – The « fontaine ardente de la Gua[13] » in France results of a leak of biogenic gas.
source : http://www.geocaching.com/geocache/GC1FH2H_la-fontaine ardente

once they have digested it, give off a natural gas called "**biogenic gas**[14]". As these bacteria are still highly active and consequently energy-needy between 35° and 45°C, it is at shallow depths of between 1 km and 1.5 km that biogenic gas will essentially occur. It forms at shallow depths and can escape to the surface to generate the well-known will o'the wisps. Eternal flames like that of La fontaine ardente (**Figure 3**) are simply natural biogenic gas leaks.

However, the bacteria will only consume a small part of the initial organic matter. The solid residue, known as kerogen, is the raw material which, at greater depths, will enable oil and gas to be manufactured in large quantities. Yet this time, it is by cooking (i.e. by increasing the temperature), that the transformation will occur.

For higher temperatures of between 80°C and 150°C i.e. for depths of 2.5 km to 5 km the long carbon chains of kerogen will undergo thermal cracking. Since the carbon content decreases and hydrogen content increases as the temperature rises, first the cracking of thick oil occurs (bitumen) and then viscous oil (heavy oils) and finally oil that is more and more fluid until, when temperatures of around 150°C are reached, it is almost petrol.

13. http://fr.wikipedia.org/wiki/Fontaine_ardente

14. In Greek bio=life, gene = engendered. Biogenic gas = gas engendered by living organisms. They can be found at the surface of the water and are called will o'the wisps.

Above 150°C, and at depths greater than 5 km, the organic matter will continue to be cracked to ultimately become natural gas. The latter, whose scientific name is *methane*, contains just one carbon atom for 4 hydrogen atoms (CH4).

Over 230°C (i.e. at depths greater than 7.5 km), the temperature becomes too high and, residual kerogen will be burned. The temperature range in which oil and gas are manufactured by thermal cracking is therefore between 80°C and 230°C. These hydrocarbons are called "**thermogenic oil and gas**[15]".

The temperature reached depends on the burial depth, which in turn depends on the accumulation rate of the organic mud. The temperature and depth scales can therefore be converted easily into time scales. Assuming a constant rate, it would take:

- Just a few million years to make biogenic gas.
- 10 to 15 million years to start to manufacture oil.
- 15 to 25 million years to start to manufacture gas.

It is also interesting at this stage to compare the consistency of the different products produced by thermal cracking and their direct relation with their carbon and hydrogen content (**Figure 4**). When the temperature increases, solid organic matter cracks and becomes enriched in hydrogen. At the same time, it liquefies, from the thick (bitumen) stage to viscous matter (heavy oils), and then liquid (light oil close to petrol) and finally gaseous (natural gas). The final solid residue which is very rich in carbon, is simply coal, distributed in the rock in the form of solid micro-clusters.

Real coal seams can only be formed however when based on a rock whose organic matter content is considerably higher than its mineral content. To ensure this, large quantities of vegetable debris (from large trees and plants) must accumulate within a shallow water layer called a peat bog[16]. As the temperature increases, a burial and baking process, very similar to that for oil and gas described previously, leads to the formation of compounds with an increasingly higher carbon content:

15. In Greek, thermo=temperature, genesis = produced. Thermogenic oil and gas = produced by temperature.

16. http://www.manicore.com/documentation/petrole/formation_petrole.html

peat (50 to 55 %), lignite (55 to 75 %), coal (75 to 90 %) and finally anthracite (> 90 %). But the hydrogen present in organic matter also generates gas called coal bed methane[17].

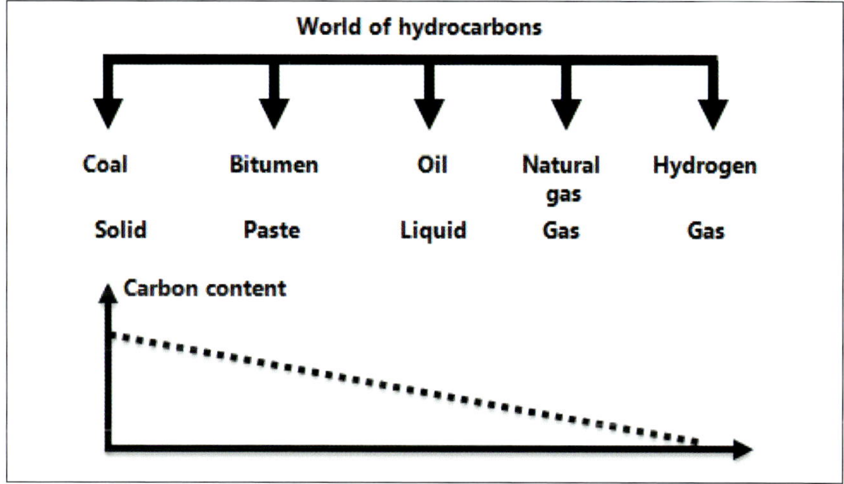

Figure 4 – The world of hydrocarbons.
A mix of carbon and hydrogen between a solid state (coal) and a gaseous state (hydrogen).

17. *Gaz de houille* in French.

What is the difference between shale oil and gas and conventional oil and gas?

The rock in which hydrocarbons form is called the *source rock.* For the *thermal cracking* recipe to be successfully completed, the source rock must be able to retain and therefore preserve organic matter and its by-products over several tens or even hundreds of millions of years. The property which enables (or prevents) a rock from retaining matter is called *permeability.* The lower the permeability of a rock, the longer the organic matter is retained inside it and preserved. Source rocks must therefore have extremely low permeabilities to ensure that the organic matter remains within them for tens of millions of years.

Despite this very low permeability, over periods of several tens of millions of years, the oil and gas end up escaping (this is called expulsion) and migrate towards the surface through much more permeable formations. If nothing stops it, migration continues right up to the surface, but if the hydrocarbons meet an impermeable formation, they become trapped in a reservoir. Hydrocarbons can travel over several dozen kilometers before becoming trapped

In short, there are two main types of hydrocarbons: the *autochthons*, which have not yet migrated from their source location (i.e. the source rock) and the *allochthons* which, after migrating from the source rock, have been trapped in a reservoir.

Reservoir hydrocarbons are called *conventional resources*. These are the resources that have been almost exclusively produced since the beginning of the XXth century. The "reservoir" rocks usually have high permeabilities. However, as the reservoir is restricted by the trap, its horizontal extension is limited to a few hundred km^2. High permeability coupled with the small extension of the reservoirs makes these "*nuggets*" difficult to find. Autochthon hydrocarbons from a source rock are called *unconventional resources or shale oil and gas*. They have astonishingly low permeabilities ranging from 1,000 to 100000 times lower than the poorest reservoirs.

However, as the source rock is not at all restricted by the presence of a trap, the extension of the impregnated zone bears no comparison to that of reservoirs.

The rock in which hydrocarbons are formed by the thermal cracking of organic matter[1] is called the *source rock*. For the recipe to be successfully completed, the source rock must be able to retain and therefore preserve organic matter and its by-products (oil and gas) over several tens or even hundred of millions of years.

The property which enables (or prevents) a rock from retaining matter is called *permeability*. The lower the permeability of a rock, the slower the flow through it of a fluid under pressure and therefore the longer the latter is retained and preserved. If organic matter and hydrocarbons in the course of formation are to be retained for millions or even tens of millions of years, the source rocks must have very low permeabilities. For this phenomenon to occur, and in keeping with Darcy's experiment[2], the microscopic clusters of organic matter are covered in a mineral mud made of very fine particles of clay, limestone and sand which, at great depths, is transformed into a hard, vitreous rock called shale (**Figure 1**).

Despite the low permeability of the source rock however, the hydrocarbons cannot be retained within it forever. Under the effect of the considerable weight[3] of the overlying sediments that have built up over tens of millions of years, the pressure of the oil and gas held in the porosity matrix[4] of the source rock increases significantly, causing the rock to crack. By seeping through these cracks, the oil and gas end up escaping (a process called *expulsion*) from the source rock and penetrate into the overlying formations which are often much more permeable (**Figure 2 – left**). Since most of these rocks are waterlogged, the hydrocarbons, which are lighter than water, migrate upwards and, if nothing stops them in their tracks, the migration continues to the surface. The presence of oil at the surface as in the *Puy de la Poix* in Auvergne (**Figure 3**) is visible evidence of the continuous migration process. Moreover, it is from the

1. This phenomenon is also called *maturing*.
2. See Appendix 3.
3. At a depth of 4 km, the sediments exert an equal pressure on the source rock... of more than a thousand times the atmospheric pressure.
4. See Appendix 4.

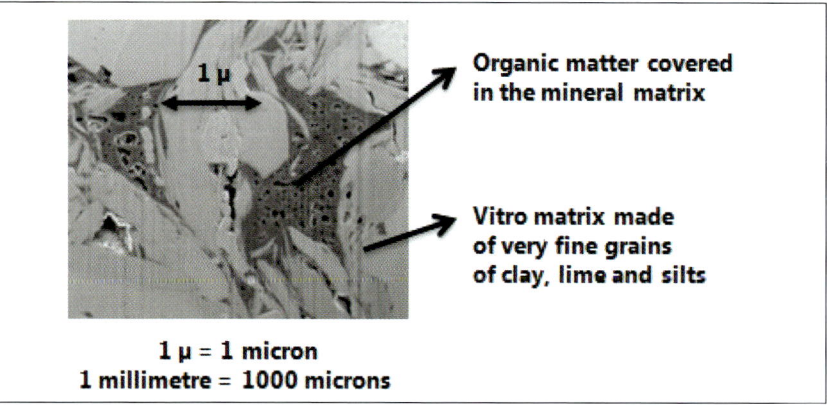

Figure 1 – Photograph taken with an electronic microscope of a cross-section of source rock showing the organic matter covered by a shaley matrix.

presence of hydrocarbons at the surface like in Pechelbronn in the Alsace region or the region of Baku in Azerbaijan, that oil exploration initially developed. People concluded that if the oil flowed up to the surface, then it must form at depth and then migrate through the geological layers.

But if by chance the hydrocarbons in their vertical migration meet an impermeable formation such as a clay or salt layer for example, as they can no longer travel upward, they will travel along the permeable layer until they reach a dome[5] where the dip[6] of the layer is inversed. As they cannot travel downward (gravity prevailing!), the hydrocarbons are trapped at the top of the structure which becomes a *reservoir* (**Figure 2 – right**). Gravity will play its role to the full by segregating the lighter gas at the top of the anticline, the oil in the middle, and the higher-density water at the bottom.

A lateral migration over about ten kilometers before encountering a trap is common. In certain cases, the hydrocarbons can cover several hundred kilometers from the source rock where they were formed before becoming trapped in the reservoir.

However, as we have seen above, migration does not occur without loss significant percentage of the hydrocarbons generated in the source rock

5. In geology this type of dome-shaped structure is called an "anticline".
6. The dip is the inclination of the layer relative to the horizontal.

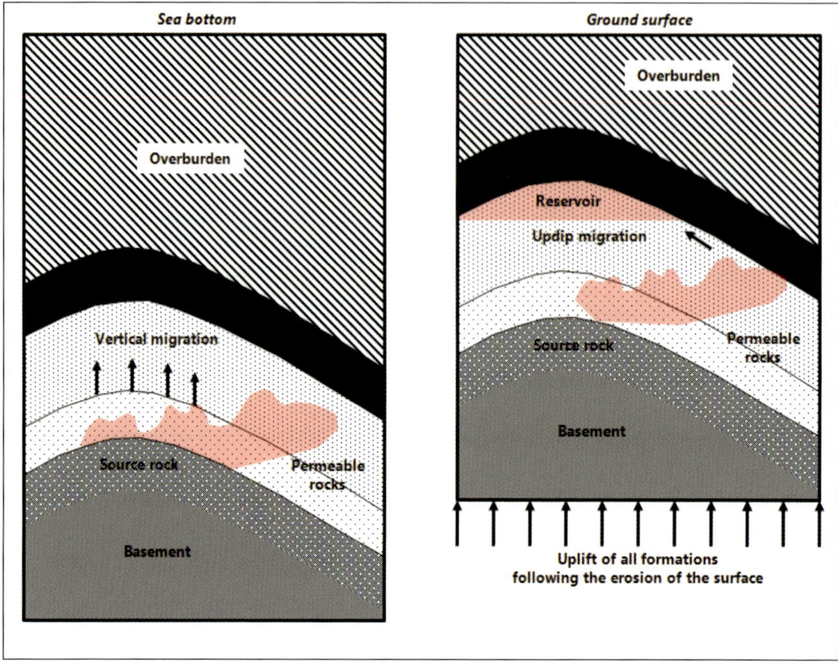

Figure 2 – Left: expulsion of hydrocarbons from the source rock and migration through the overlying permeable layers.
Right: Hydrocarbons migrating along an impermeable layer before becoming trapped in a reservoir.
Uplift subsequent to the erosion of part of the overburden.

and which do not encounter any traps, migrate to the surface and oxidize in the air. The yield (percentage of hydrocarbons trapped in the reservoir compared with those that formed in the source rock) is highly variable. Even though in certain exceptional cases, the yield can be higher than 50%, in numerous basins it is lower than 1%. The hydrocarbons trapped in the reservoir therefore only represent a small fraction of the hydrocarbons that are made in the source rock.

The final geological phenomenon that is important to consider is the erosion[7] of surface formations (also called overburden - **Figure 2**) when

7. Erosion is the process of rock degradation caused by a series of external physical factors such as water, wind, or frost, and also chemical (e.g. karst dissolution).

Figure 3 – Surface bitumen (Puy de la Poix – Auvergne)

they emerge at sea level following, for example a withdrawal[8] of the sea or an uplift subsequent to the formation of a mountain range. In a few million years, rock can be eroded by several kilometers. So a source rock that was initially located at a depth between 5km to 7km when it was forming hydrocarbons may find itself today at a depth of 2km to 4km.

In short, there are two main types of hydrocarbons: the *autochthons*, which have not yet migrated from their source location (i.e. the source rock) and the *allochthons* which, after migrating from the source rock, have been trapped in a reservoir.

Allochthon reservoir hydrocarbons are also called conventional resources because they are the resources that have been produced almost exclusively since the beginning of the XXth century. The reservoir rocks are almost always sandstone (consolidated sands) or limestone with a large granulometry, the high permeability of which is between 1 Darcy (excellent permeability) and 1 milliDarcy[9] (reputedly poor permeability). However, insofar as the reservoir is constrained by the trap (i.e. by the crest of the anticline), the horizontal extension of the zone impregnated by hydrocarbon is restricted to a few hundred km^2 or even, in exceptional cases, a few thousand km^2. The giant Kashagan field in the Caspian Sea covers a surface area of 3,000 km^2 (equivalent to Greater London) and the Gawhar field in Saudi Arabia recognized as the most extensive field in history, covers just over 5,000 km^2. The high permeability but small extension of the reservoirs makes these nuggets difficult to find.

8. In geology, this phenomenon is called *"marine regression"*.

9. 1 milliDarcy is a thousandth of a Darcy.

Autochthon hydrocarbons from a source rock are also called *unconventional resources* or *shale oil and gas*. Although knowledge of their existence dates back a long way, it was only very recently that they began to be developed massively (since 2005) in the United States and to a lesser extent in Canada. A mixture of very fine grains of clay, limestone and sand[10], shales have astonishingly low permeabilities ranging between 10 and 500 nanoDarcy[11] that is 2,000 to 100,000 times lower than those in the poorest reservoirs. However, as the source rock is not at all constrained by the presence of a trap, the reach of the impregnated zones bears no comparison to that of reservoirs (**Figure 4**). The Barnett formation in Texas is comparable to the surface of Belgium (30,000 km^2), the Marcellus in Pennsylvania covers the equivalent of half of France (250,000 km^2) and the Bazhenov in Western Siberia, just short of a million km^2 (i.e. almost twice the size of France).

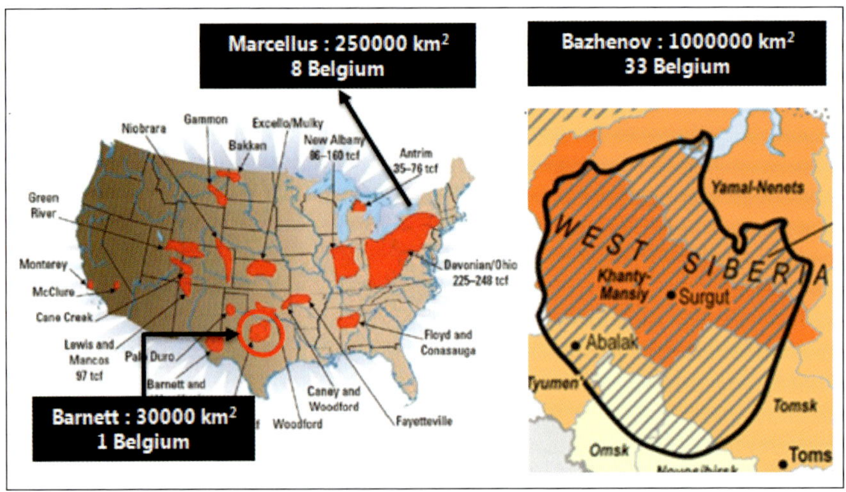

Figure 4 – Compared to reservoirs, source rocks extend over thousands of square km. Barnett is equivalent to the surface area of Belgium, while the Bazhenov extends over nearly two times the surface area of France.

10. The specific term for very fine grains of sand is "silt".

11. A nanodarcy is equal to a billionth of a Darcy.

How can subsurface oil and gas resources be accessed and brought to the surface?

In order for the hydrocarbons concentrated in the subsurface (whether in a source rock or a reservoir) to be used, they must be accessed, then collected and brought to the surface. It was the invention of rotary drilling, which consists of rotating a drill bit at the bottom of the borehole to break up the rock, which was to pave the way for the modern oil industry. Given the heterogeneity of the layers encountered, the well has to be drilled in several sections. Once they have been drilled, they are covered by a steel tube (called casing) which is cemented in place. At the depth of the oil or gas layer the casing is perforated to allow the fluid to penetrate the well and flow to the surface by means of its natural pressure. The assembly of the cemented casing creates a seal which prevents the hydrocarbons from coming into contact with the layers they pass through, in particular the water-bearing surface layers.

The first onshore fields were produced using a large number of vertical wells clustered together. Offshore however, the *cluster* concept had to be invented to minimize the number of platforms. This consists of drilling several dozen wells from a single platform and then deviating the well trajectory to reach the impregnated rock, sometimes several kilometers away. To drill a deviated well, the end of the drill string needs to be bent slightly to give the borehole a curved shape in terms of both the inclination relative to the vertical, but also in terms of azimuth relative to the North. For a horizontal well, the aim is to *land* in the impregnated rock at an angle of 90°, then continue drilling horizontally over several hundred or several thousand meters.

Apart from giving access to distant targets, deviated and particularly horizontal wells significantly increase the surface of the well exposed to oil or gas as the distance of rock drilled through is much longer. For horizontal wells the horizontal component can sometimes represent 20 to 40 times the vertical thickness of the layer, or even more, so this type of well is particularly suitable for low-permeability rocks.

However, even though horizontal wells partially compensate for the permeability deficit between a very good reservoir and a poor reservoir, it is not at all sufficient for exploiting unconventional resources, where permeabilities are 2,000 to 100,000 times lower than that of a poor reservoir.

Identifying the subsurface rock which contains oil or gas (whether it is a source rock or a reservoir) is the first crucial step. However, for the energy concentrated in it to be used to heat a building, drive a car or produce electricity, the hydrocarbons must be accessed, collected and brought back to the surface and then ultimately transported to the site where they will be directly consumed or transformed into electricity.

A little history

"If you can't get to the oil, let the oil to you" would be a succinct way of describing the beginnings of oil production. Hydrocarbons that form deep in the subsurface and do not meet any trap, migrate to the surface and then seep through rocky outcrops, produced the first oil fields. The shores of the Caspian Sea are thought to be the cradle of the oil industry because, during the IXth century, oil is thought to have been produced in the Baku region of Azerbaijan. Marco Polo who visited the Azerbaijani capital in 1264 described *"seepages from which oil flows abundantly"*. At the end of the XVIth century, trenches 35 meters deep were dug to collect the seep oil, but the first great industrial adventure was the production of bitumen in Péchelbronn, Alsace. The primitive production technique consisted of digging a trench 3.5m deep around the source, letting the oil seep to the surface and then collecting it.

To access seep oils that were a little deeper, a mining technique developed, which consisted of digging underground galleries. Unfortunately production was often delayed by accidents and, in many cases, people had to resign themselves to abandoning the flooded galleries as the pumps at that time were not powerful enough to extract the water. Not only that, but at a certain depth gas pockets would continuously appear, causing fatal explosions.

The increasing demand for oil in the first half of the XIXth century could not be met by the low irregular supply of seep oil. If there were surface seepages, they were a sign that the hydrocarbons were formed at

Figure 1 – Colonel Drake's oil well in Old Creek near Titusville, Pennsylvania.

a greater depth and then percolated to the surface. Gaining direct access to deep-lying hydrocarbons through a well appeared to be the solution for the future. The most historic date is still that of August 27, 1859 when *Colonel Drake* dug a well using a drill bit suspended by a cable and driven by a steam engine (**Figure 1**). Having drilled to a depth of just 23 meters, he succeeded in bringing the black gold to the surface at the unprecedented rate of 10 barrels[1] per day. However, the limitations of the percussion technique soon came to light: slow penetration rates, the need to stop drilling to extract debris and major constraints in terms of depth. It was the discovery of rotary drilling, which consisted of rotating a bottomhole tool designed to crush the rock via a metal drill string, which really laid the foundations of the modern oil industry.

Rotary drilling is thought to have been used since ancient times but, despite numerous projects (one of which is attributed to Leonardo da Vinci), it did not gain any real popularity until the second half of the XIXth century. The idea[2] was to pump a fluid inside a metal pipe to carry the debris up to the surface through the annular space (**Figure 2**). In the end, it was only at the beginning of the XXth century that the first rotary

1. One barrel equals 159 liters.
2. The patent of the Englishman Robert Beart was based on an idea from a Frenchman called Flauville.

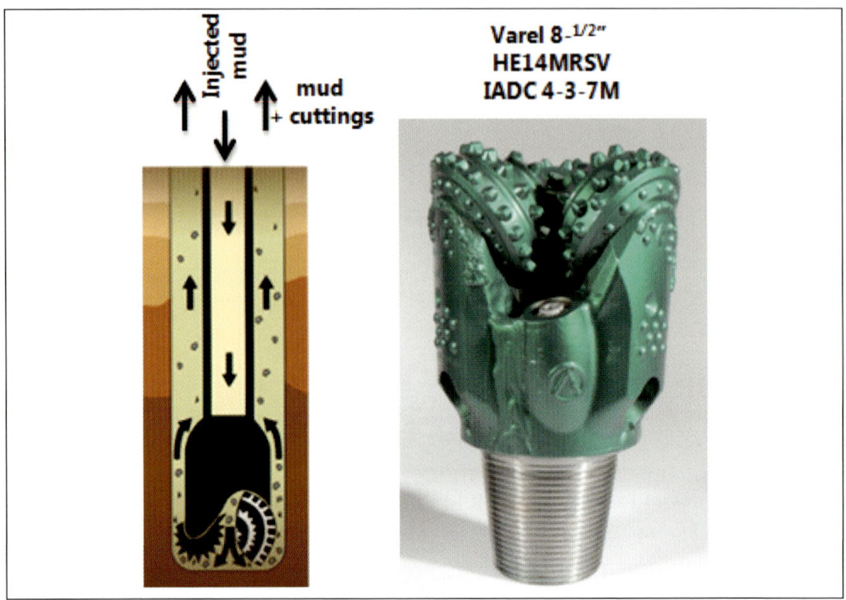

Figure 2 – Rotary drilling and example of a drill bit (courtesy of Varel)

mode oil wells were drilled in soft rock in Texas and California. The invention of the steel drill bit[3], which crushes the rock into little chips[4], allowed for drilling through increasingly hard formations at ever-increasing depths, as early on as the 1920s.

The well

Because of the heterogeneity of the layers drilled through (differences in terms of the hardness of the rock and of the pressure of the fluids) between the surface and the oil or gas-impregnated rock (whether reservoir or source rock), it is impossible to drill a well of uniform diameter. It must be drilled in several sections and, once completed, the sections are covered by a steel tube (called *casing*) which is cemented in place. When finished, a well is therefore a telescopic structure, starting out with a diameter of about 80 cm, shrinking to about 10 cm when it finally reaches the oil or

3. The drill bit is a real crushing tool made up of three toothed wheels. To begin with, the teeth were steel spikes, but were later replaced by harder materials such as tungsten carbide and diamond that could cut through the hardest rocks.
4. These chips of rock are called *cuttings*.

gas-impregnated rock. A typical well will include on average four sections of cemented casing. At the level of the oil or gas layer the casing is perforated to allow fluid to flow into the well and rise to the surface under the action of its natural pressure inside uncemented tubing (**Figure 3**).

The assembly of the cemented casing sections creates a seal which prevents the hydrocarbons from coming into contact with (and therefore contaminating) the different formations through which they pass, in particular the water-bearing surface layers. This is known as *well integrity*.

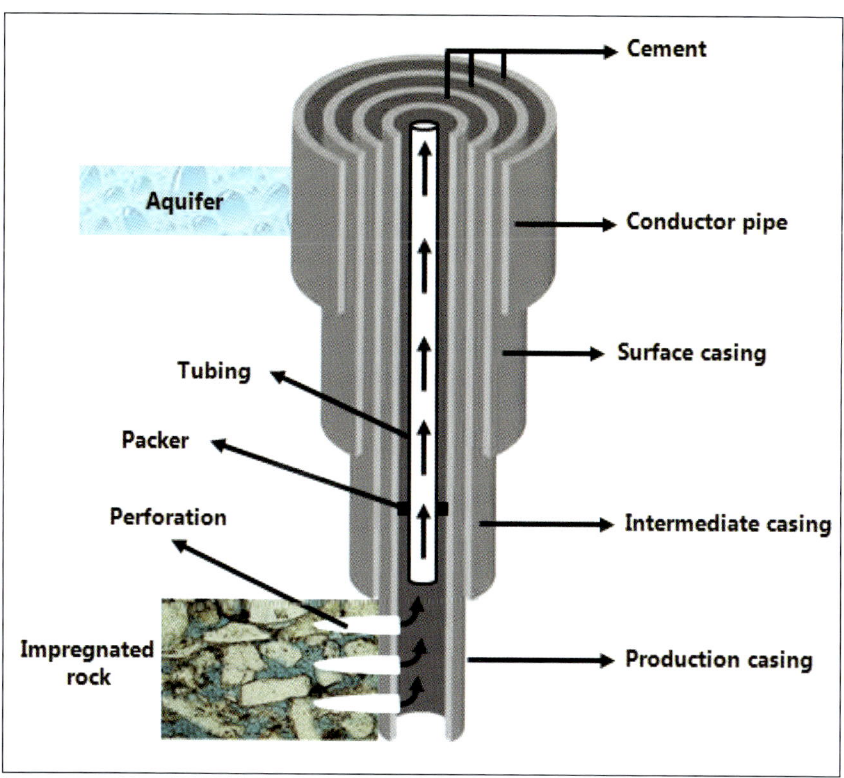

Figure 3 – Typical well architecture. The cemented casing assures the well seal and protects the water-bearing surface layers. The production fluid flows up through the tubing

The first vertical developments

The first onshore fields[5] were produced using a large number of vertical wells clustered together and located in the immediate vicinity of the surface seeps (**Figure 4**). By highlighting the concept of an anticline trap, geologists added some logic into well location techniques and contributed to rationalizing the number of wells drilled. However, it was the discovery of oil fields in the North Sea and in the Gulf of Mexico in the mid-1970s that really revolutionized drilling techniques, moving first toward deviated and then horizontal wells.

Deviated and horizontal wells

In order to develop offshore oil and gas fields, the number of platforms must be reduced. This constraint gave rise to the *cluster* concept which consists of drilling several dozen wells from a single platform and then

Figure 4 – Typical examples of production using vertical wells in the first half of the XXth century.
1 Baku region (Azerbaijan). – 2 and 3 California. – 4 Maracaibo Lake.

5. Except for the case of the shallow waters of Lake Maracaibo in Venezuela.

deviating the trajectory of the borehole in terms of both inclination (angle of the well axis relative to the vertical) and in terms of azimuth (angle of the trajectory relative to geographical North).

From points very close together at the surface (the start of the boreholes are just two to three meters apart), this technique means that the oil- or gas-impregnated rock can be reached at distances (this horizontal distance is called departure) of several km (**Figure 5**).

Controlling the borehole trajectory means first having a specific drill string assembly which can create the required deviation. This is the role of the directional downhole motor (**Figure 6**) comprising a helical screw attached to the end of the drill string and driving the rotation of the drill bit using the movement of the drilling mud[6]. By slightly bending (one to three degrees on average) the end of the drill string assembly between the motor and the drill bit, the borehole is drilled in a curve and its trajectory can be selectively deviated in terms of both inclination and azimuth. A well trajectory would therefore turn progressively by a few degrees every 100 meters.

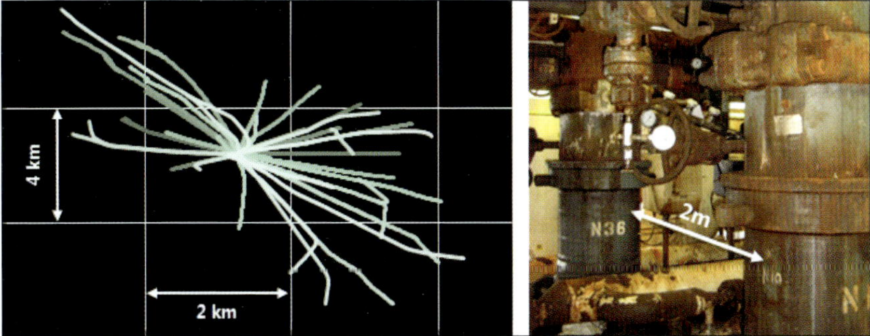

Figure 5 – Example of a cluster in the North Sea. Left: plan view. Many wells are deviated in different directions. Right: the start of two wells just two meters apart at the surface

6. Contrary to traditional rotary drilling, this technique does not require the drill pipes to be rotated.

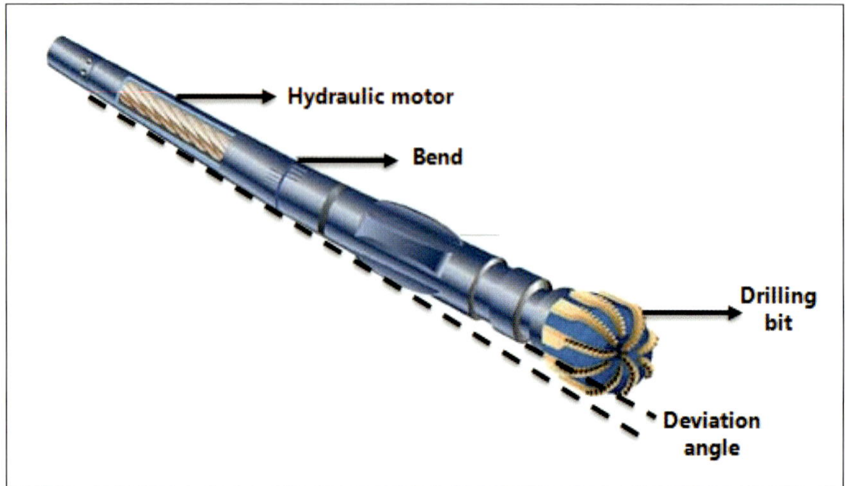

Figure 6 – Directional downhole motor

Using this technique different shapes of borehole can be drilled, the most common being the **J-shape** which tends toward a continuous inclination, the **S-shape** which, once the borehole has been inclined, aims to bring the trajectory back to the vertical and finally the **H-shape**, where the aim is to "*land*" in the impregnated rock at an angle of 90° and then continue drilling horizontally for several hundred or even thousand meters. The well inclination and azimuth are estimated constantly using sensitive measuring instruments which are an integral part of the drill string. Apart from the possibility of accessing distant targets, deviated and particularly horizontal wells, considerably increase the surface of the borehole exposed to the oil and gas as the distance of rock drilled through is markedly longer. So for a given oil or gas layer thickness, a well inclined at 70° will double the length of a vertical well (**Figure 7**). For horizontal wells, the linear portion can sometimes be 20 to 40 times the vertical height of the reservoir, if not more.

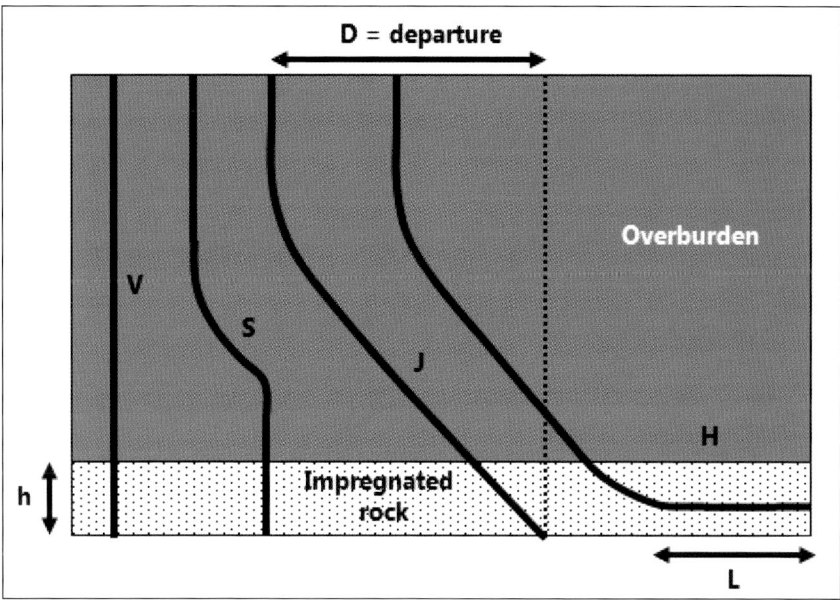

Figure 7 – Different well trajectories (V= vertical, S-shape, J=inclined, H=horizontal). Compared with the vertical borehole, the inclination creates an offset, but more importantly increases the borehole surface area exposed to the impregnated rock.

Producing a well

The production of a well is defined as the flow rate (number of cubic meters per hour) at which the oil and gas flow through the rock and up the well to the surface. To highlight the main parameters that govern the production of a well, the Darcy's law may be taken as a starting point (see **Appendix 3**). In the diagram shown in **Figure 8**, A is the exposure surface between the borehole and the rock, i.e. for the same borehole diameter, the length of rock through which it passes.

According to Darcy's law, the production of a well is directly proportional to the product ***k*A*** (rock permeability multiplied by the length of rock through which the borehole passes).

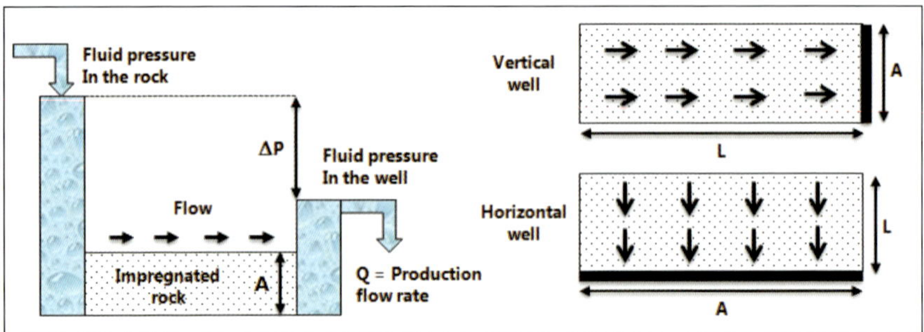

Figure 8 – Darcy's law.

Let's take three rocks impregnated with the same fluid in a layer 10m thick but with different permeabilities (**Figure 9**): an excellent reservoir of 100 mD, a poor reservoir of 1mD and finally a source rock of 100nD. Let's start by drilling a vertical well into each of the three layers. The exposed length is therefore equal to the thickness of the layer, that is 10 meters. To keep things simple, let's assume that the good reservoir produces 1,000 m^3 of oil per day via the vertical well. For the same exposure, the poor reservoir will produce only 10 m^3/day. As for the source rock, the very low permeability means that it will produce barely one liter of oil per day.

Now let's consider drilling a horizontal well 1,000 meters long into the poor reservoir and the source rock. Thanks to the significant increase in the length of rock exposed, this time the poor reservoir manages to produce as much as the good reservoir, i.e. 1,000 m^3/day. However, production from the source rock is still a mediocre 100 liters per day. To compete with the 10 m^3/day of the vertical well in the poor reservoir, the source rock would have had to be exposed over 100km of horizontal borehole and, to reach the production rate of the good reservoir (1000 m^3/day), the horizontal borehole would have had to be an incredible 10,000km long, i.e. the distance from Paris to Tokyo.

An horizontal borehole can therefore compensate for the low permeability of a poor reservoir. However, it is largely insufficient to make the production of a source rock economically viable. To achieve this goal we must play with the second key parameter of the product ***k*A***: permeability. So for example, by multiplying the permeability of the source rock by 100, a horizontal borehole 1,000m long would be able to

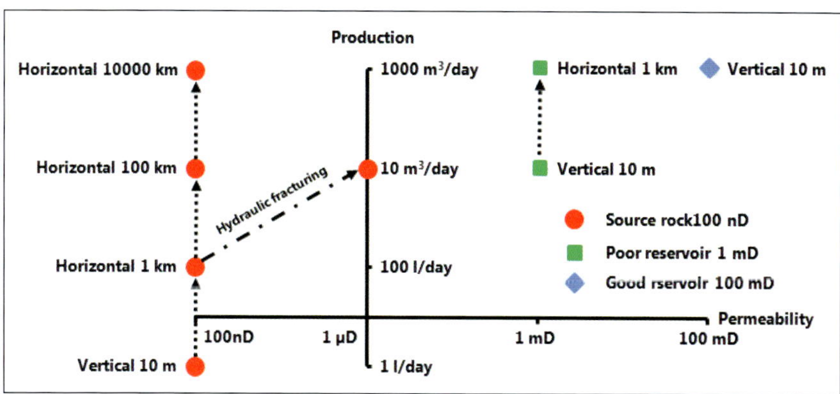

Figure 9 – Comparison of the production rates of a good reservoir of 100mD with that of a poor reservoir of 1mD and a source rock of 100nD. To attain a mediocre production rate of 10m³ per day in the source rock, a horizontal borehole 100 km long would have to be drilled. Horizontal drilling technologies are not therefore sufficient to make the production of shale oil and gas economically viable.

produce 10 m³/day (production identical to that of the vertical borehole in the poor reservoir), and, by drilling 3,000m of horizontal borehole (which is technically possible), the production rate would reach 30 m³/day, i.e. approximately 200 bbl/day.

The method which artificially increases the permeability of a rock exists: it is the famous **hydraulic fracturing** technique.

Hydraulic fracturing: why is it necessary to fracture the source rock to extract shale oil and gas?

Hydraulic fracturing is a mature technology. The first pilot dates back to 1947 and, since this historical premier, over 2.5 million hydraulic fracturing operations have been carried out. Initially used in vertical and deviated wells, the technology was applied to horizontal wells during the 1990s and it is the combination of horizontal drilling/hydraulic fracturing that enables the economically viable production of shale oil and gas.

The best analogy for understanding hydraulic fracturing is the road network. In a conventional reservoir with good permeability, the secondary roads represent the permeability of the rock matrix and the main roads represent a "*grid*" of natural fractures. The network means that the liquid can be carried easily toward the well. For an unconventional rock, nature has not laid any main roads, or secondary roads, but rather a series of poor country roads that are unsuitable for traffic. To access the resources and bring them to the well, the road network needs to be artificially created. In practice, the multi-hydraulic fracturing of a horizontal well does this job. It consists of fissuring the rock using a liquid (usually water) injected at high pressure. However, if nothing further was done, the fracture would close up under the effect of earth pressures when the injection stops. To keep the fracture open, calibrated sand or ceramic microbeads (proppants) are added to the liquid being injected to hold the fracture open when the injection process stops. The network required is created by repeating the fracturing process, tipically every 100 meters, along the horizontal well. In conventional rocks with good permeability, there are no real limits to the size of producible areas. In the case of source rocks however, it is the fineness of the network created that will govern production and only those hydrocarbons around the edges of the producible area can be exploited. This explains on the one hand, the rapid decline in well production and on the other hand, the poor overall recovery rate. Whereas for conventional hydrocarbons about a third of the oil and 70 to 80% of the gas is recovered, for unconventional hydrocarbons, the recovery rates stand at just 5 to 10% for shale oil and 15 to 20% for shale gas. The rapid decline in well production means that new wells need to be drilled and fractured continuously to maintain a satisfactory overall production level.

Hydraulic fracturing is a mature technology which significantly predates the first horizontal wells and the production of shale oil and gas. The first pilot dates back to 1947 and was run by the American company Halliburton, in the conventional Hugoton fields in Kansas (**Figure 1 – left**). Since this historical premier, over 2.5 million hydraulic fracturing operations have been carried out worldwide. Alongside the oil and gas industry the technology is routinely applied to produce geothermal energy from hot, dry rocks like those in Soultz in Alsace, France (**Figure 1 – right**). Initially used in vertical and deviated wells, the technology was applied to horizontal wells during the 1990s and it is the combination of these two mature technologies (horizontal drilling and hydraulic fracturing) that enables the economically viable production of shale oil and gas.

Figure 1 – Left: the first hydraulic fracturing experiment (Hugoton field – Kansas). Right: Soultz geothermal plant (Alsace, France)

Hydraulic fracturing: why?

In question 6, we saw that contrary to conventional hydrocarbons, horizontal drilling technologies do not **alone** allow produce shale oil and gas in sufficient (i.e. economic - **Figure 2**) quantities. Another technique ***hydraulic fracturing*** must be used in conjunction with the first, to artificially increase the permeability of the source rock.

The best analogy for understanding the advantage of hydraulic fracturing in rocks with very low permeabilities is that of the road network, used to supply a town with cereals for example **(Figure 3)**. The network comprises secondary roads that crisscross each cereal plot which in turn is connected to a network of main roads. The network (number and width of roads) must be adapted to the agricultural production of the cultivated area. If it is not sufficient, it will not be able to carry the cereal harvest to the city fast enough. If it is overly sufficient, it will not be used to its full

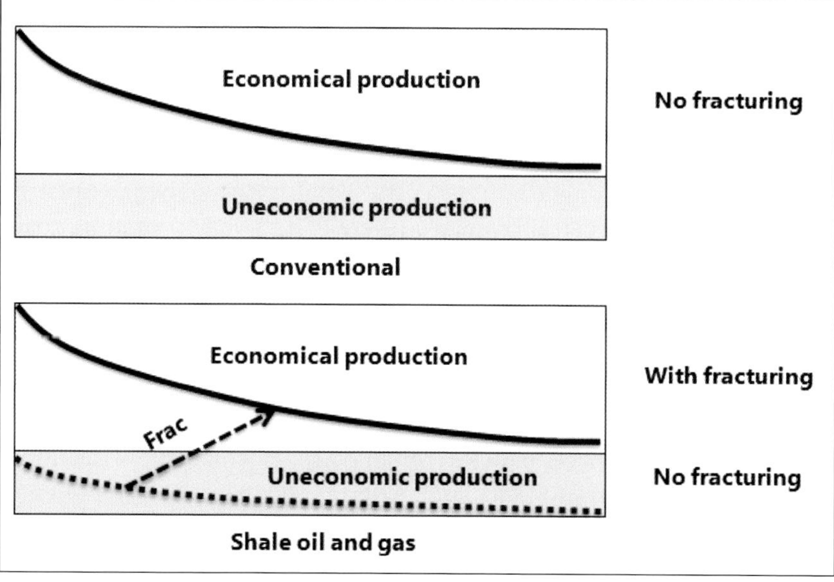

Figure 2 – Top: conventional hydrocarbons do not need hydraulic fracturing. The permeability of the rock is sufficient to allow natural, economically viable production. For shale oil and gas, natural production is not economically viable. Hydraulic fracturing brings it from the non-economical to the economical zone

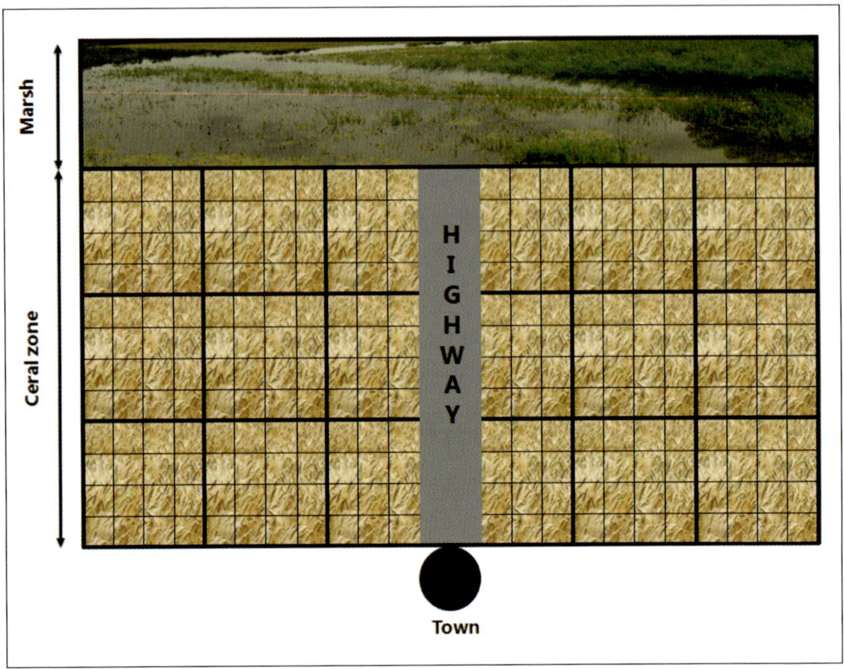

Figure 3 – Analogy of hydraulic fracturing and the road network.

capacity and will generate extra costs. Finally, the network must be confined within the cereal crop area (there is no point in creating roads in the marshy area) and lorries must be able to meet at the point where the highway joins the city without creating too many traffic jams.

Using this analogy, let's now imagine a conventional reservoir with a good permeability (somewhere between 100mD and 1D). The city represents the well, the secondary roads represent the permeability of the rock matrix (micro spaces between rough particles which allow the liquid contained in the pores to percolate), whereas the main roads represent a gridded network of natural fractures that cut through the rock. This well-connected network carries the liquid to the well at a high flow rate. To represent a poor quality reservoir (permeability somewhere between 1mD and 10mD), the network of main roads would be removed (natural fractures) which would leave just the secondary roads (matrix permeability). This would affect the liquid flow rate and therefore reduce production. In the case of conventional reservoirs, nature offers a free network of

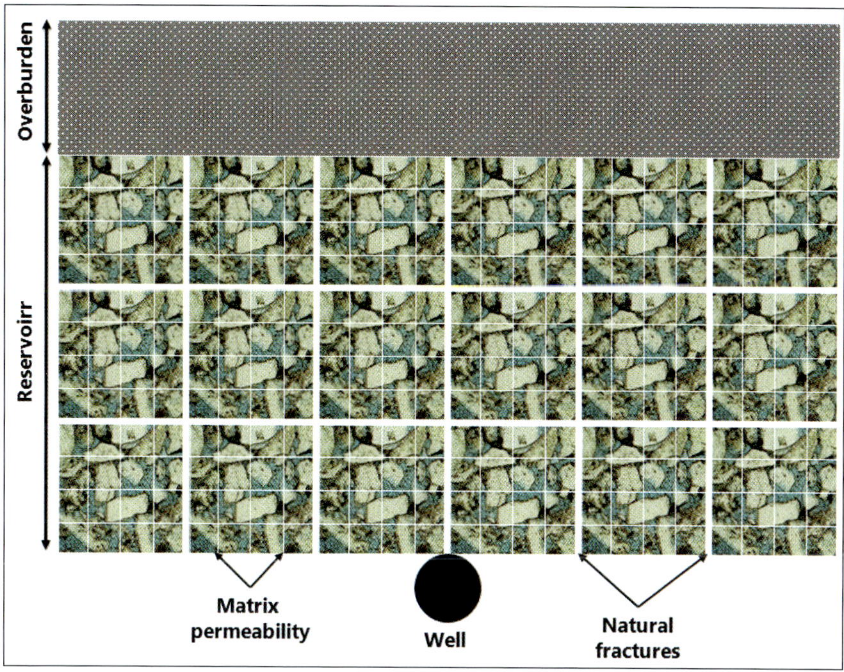

Figure 4 – Analogy of the road network in a reservoir (vertical view). The national road network represents the network of natural fractures. The network of secondary roads represents the matrix permeability (spaces between the particles where the fluid can percolate).

secondary roads, and in certain cases, a network of main roads on top of that, so there is no need to create any artificial ones.

Now let's consider the case of an unconventional rock containing shale oil and gas. The vitreous matrix has a natural permeability of a few dozen nanoDarcy. Even though shales are often naturally fractured, the fissures are often closed, blocked and usually not connected to the well. This time, nature offers neither main roads nor secondary roads, but a sketchy set of poorly-surfaced country roads unsuitable for traffic (**Figure 5 - top**). To access the resources and bring them to the well, a network of secondary roads needs to be **artificially created** by tarmacking the country roads to make them suitable for traffic, and creating new roads to connect the network of secondary roads to a highway, which is connected to the well (**Figure 5 – bottom**). In any case, the size of the road network will never be able to compete with the porous network of a conventional rock,

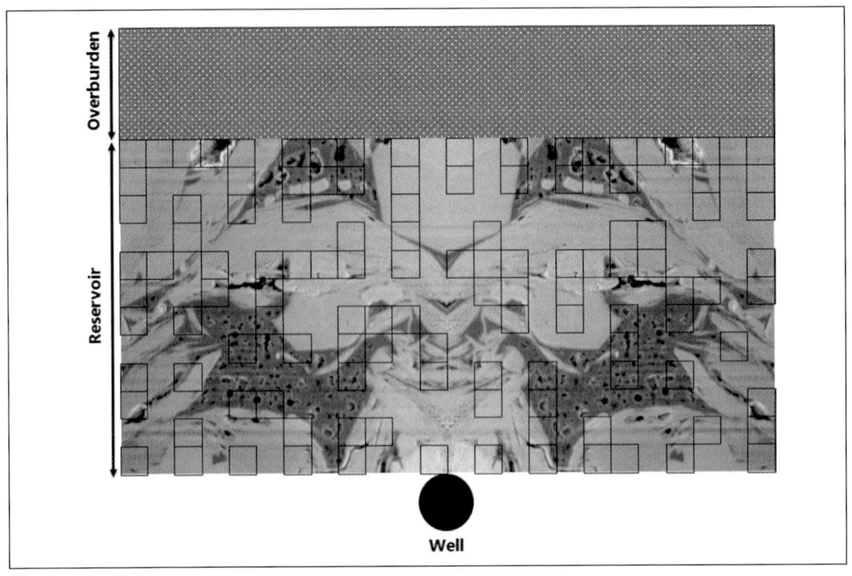

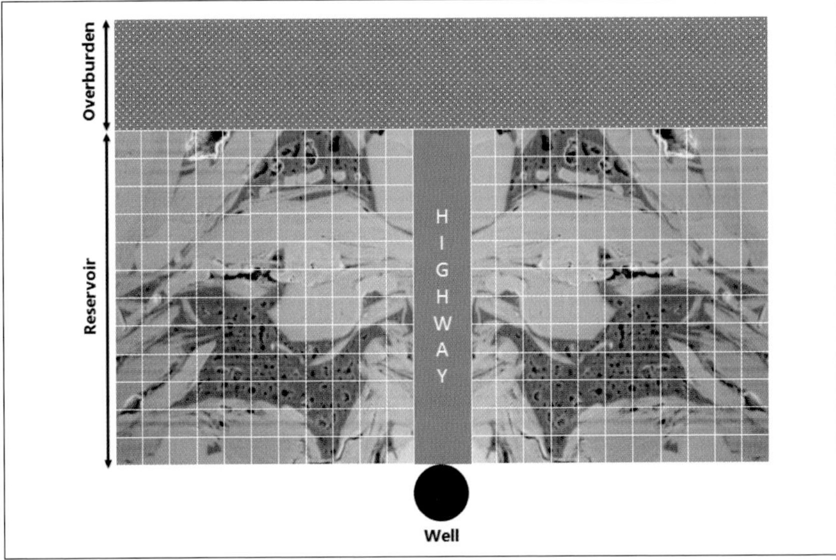

Figure 5 – Top: Source rock before fracturing. The matrix provides no secondary road network leaving the natural fissuring of disconnected country roads that are unsuitable for vehicles. Bottom: after fracturing, a network of secondary roads has been opened and connected to a highway, linked to the well.

which is governed by its particle size. Consequently, only the resources located on the roadside of the secondary roads will be accessible, whereas the resources located in the middle of the plot boundaries will remain inaccessible. This explains on the one hand the poor overall recovery rate and the extremely rapid decline in well productivity. We will come back to these two points later. The set contained within the network of highways/secondary roads is called the ***SRV*** (Stimulated Rock Volume).

How does hydraulic fracturing work?

In practice, it is the **hydraulic multi-fracturing of a horizontal well** that will create a SRV and enable the shale oil and gas to be produced. In hydraulic fracturing, the first term designates the branch of physics which studies the circulation of liquids under pressure[1], and the second term means fissuring matter under the effect of a mechanical pressure.

Hydraulic fracturing is therefore the process of fissuring a subsurface rock using a liquid (usually water) injected at high pressure from the surface into a well. Fracturing initiates when the pressure of the liquid pumped into the well exceeds the geological thrust (the exact term is stress) that pervades in the deep rock. By contrast to liquid, this thrust, which is mainly the result of the weight of the cap rock[2] and the forces from plate tectonics[3] is not equal in all directions (**Figure 6**). At great depth, the vertical thrust is generally greater than the horizontal thrusts which themselves can vary considerably. Under the effect of the hydraulic pressure of the liquid exerted at the borehole wall (for now, we will assume the well is vertical), the latter will split at the point at which nature requires the least possible effort, that is, vertically and perpendicular to the minor (horizontal) thrust. As long as the pressure of the liquid remains

1. In earlier physics, hydraulics was the science that taught how to measure, direct and lift water. Hydraulic machines were mainly the pumps used to this effect and hydraulics was therefore the remit of the pipe engineer. http://fr.wikipedia.org/wiki/Hydraulique
2. At a depth of 2,500m, the weight of the overlying land exerts a vertical pressure approximately 650 times that of atmospheric pressure.
3. The outer layer of the Earth, or "*lithosphere*" is cut up into rigid plates that float and move on the upper mantle (asthenosphere) which is more fluid and ductile. The impact of these plates against one another generates very powerful directional pressures (constraints) at depth. The main mountain chains (Alps, Pyrenees, Himalaya, Andes) were created as a result of one plate hitting another, and the most violent earthquakes are another result of the impact between plates.

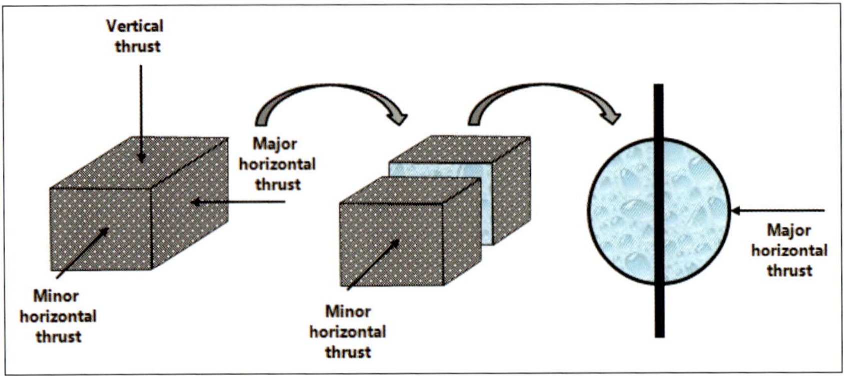

Figure 6 – Hydraulic fracturing: general concept. The geological thrust is different in each direction. The hydraulic pressure initiates perpendicular to the minor horizontal thrust. The fracture therefore propagates vertically and parallel to the major horizontal thrust.

higher than that of the minor horizontal thrust, the fracture will continue to open and propagate in a more or less circular fashion within the layer. Depending on the volume of liquid injected, the diameter of the fracture can reach several tens or exceptionally several hundred meters.

However, if nothing further done, the fracture that had been kept open solely by the pressure of the liquid would close up under the effect of the weakest horizontal thrust once the injection process stops. To keep the fracture open when the pressure drops, graded sand or ceramic beads are added to the water injected. This *"propping agent"* will line the fracture, retain it and keep it open once the injection process stops (**Figure 7**).

However, hydraulic fracturing in a vertical well opens just one vertical fracture, generally a single highway in the analogy of the road network. We are therefore still a long way short of having created a connected network of highways and secondary roads. The combination of hydraulic fracturing and the horizontal well is the key to achieving this objective.

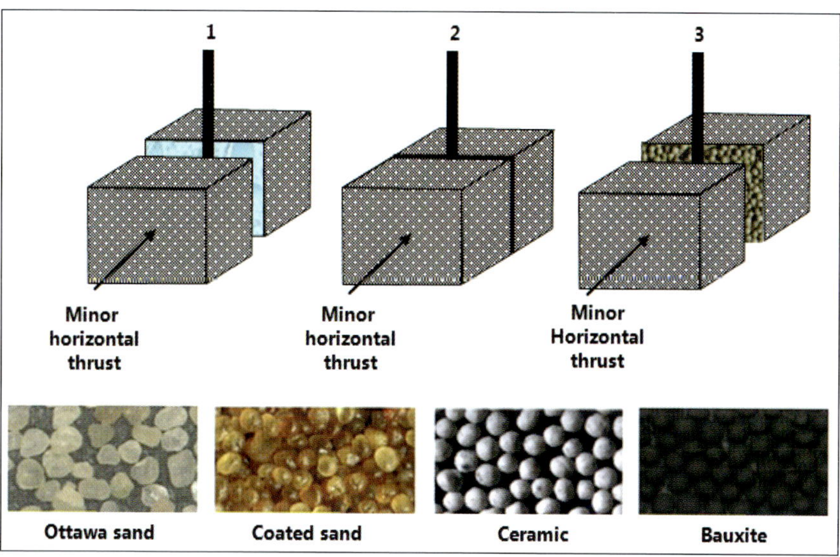

Figure 7 – Top: propping a hydraulic fracture. 1 – During injection, the fracture is held open by the pressure of the liquid. 2 – Without further precautions, the fracture would close up once the injection process stops. 3 – The propping agent injected with the liquid supports the fracture and keeps it open in spite of the weakest horizontal thrust. 4 – Bottom: example of proppants used in the oil industry.

Creating a SRV by combining a horizontal well and hydraulic multi-fracturing

Contrary to the vertical well which is unique, the horizontal well can be drilled in different directions. Two specific cases are shown in **Figure 8**. If the well is drilled in the direction of the major horizontal thrust, it will fracture parallel to the axis of the well and a single longitudinal fracture will be created. However, if the well is drilled in the direction of the minor horizontal thrust, the fracture will be initiated perpendicular to the axis of the well. By repeating this process several times along the borehole, starting from the furthest end, a series of transverse fractures can be created. Each of the fractures is called a fracturing stage. In shale oil and gas formations, the process is usually repeated every 100 meters. So for a 1 to 2 km horizontal well, between 10 and 20 fracturing stages will be performed.

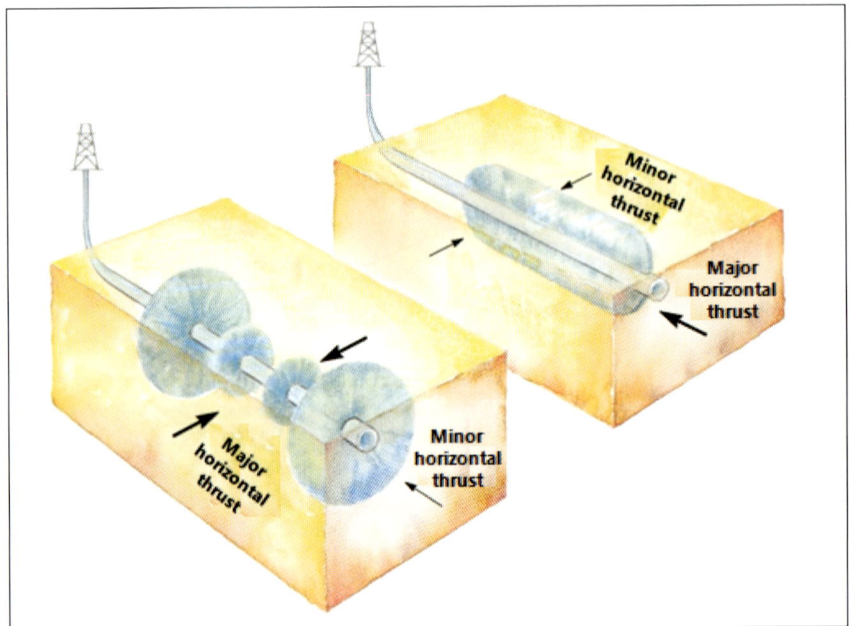

Figure 8 – Hydraulic fracturing in a horizontal well depends on the borehole direction. Top: the well is drilled along the direction of the major horizontal thrust. A single longitudinal fracture is created. Bottom: the well is drilled in the direction of the minor horizontal thrust. It is then possible to generate a series of transverse fractures perpendicular to the well.

Thanks to these transverse fractures propped by sand we were able to create a series of highways within the rock. All that remains to be done now is to build a network of fissures between the main fractures that is fine enough to recover the hydrocarbons located on the plot boundaries (**Figure 9**). The network of secondary fissures is built on the existing natural network of fissures (the country roads unsuitable for vehicles) that are reopened and propped by injecting sand. However, to complete the secondary road network, new fissures must also be created and propped open so that the entire set can be connected to the highways. The fissured area, delimited by the main transverse fractures is the SRV where the equivalent permeability will be 100 to 1,000 times greater than that of the initial rock.

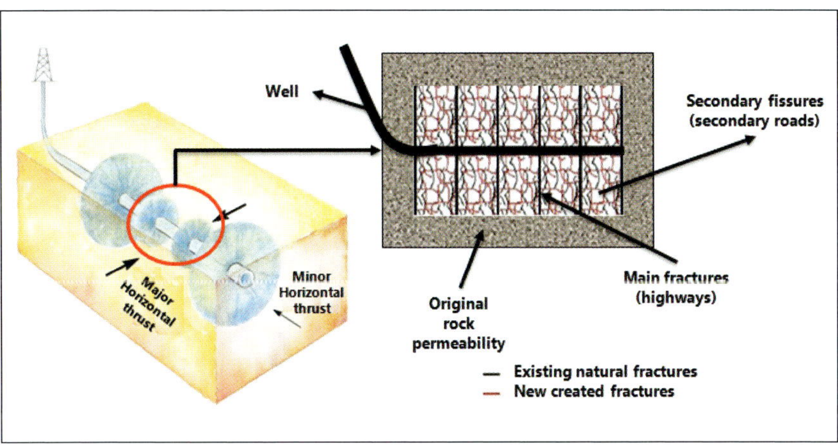

Figure 9 – Constructing the SRV. Between the main fractures, the existing natural fissuring is reactivated and propped. New fissures are also created. The fissured network then connects up with the main fractures.

In spite of all these efforts, the decline in well production is extremely rapid

In conventional rocks with a good permeability, there are no real limits to the size of producible areas. In the case of source rocks however, it is the fineness and the connectivity of the SRV which will control the initial well production, and the rapidity of its decline. Only those hydrocarbons located on the edges of the plot and connected to the SRV can be produced (**Figure 10**). This explains on the one hand, the rapid decline in well production and on the other hand, the poor overall recovery rate. Whereas for conventional hydrocarbons about a third of the oil and 70 to 80% of the gas contained in the reservoir rock is recovered, in the case of unconventional hydrocarbons, the recovery rates stand at just 5 to 10% for shale oil and 15 to 20% for shale gas.

The rapid decline in well production means that new wells need to be drilled and fractured continuously to maintain a satisfactory overall production level. Later, we will come back to this frantic race for drilling and fracturing.

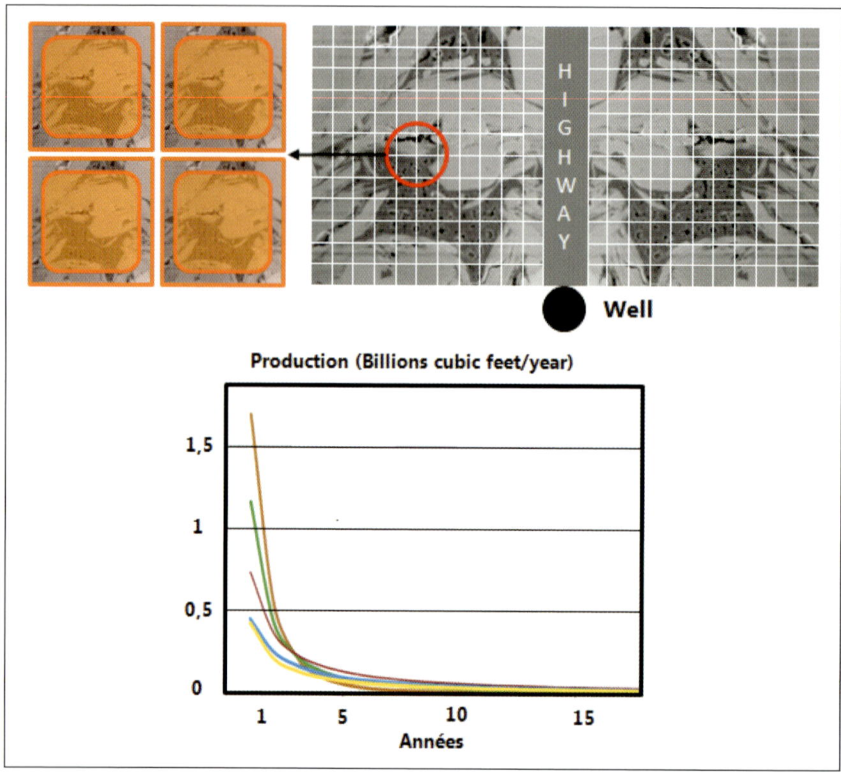

Figure 10 – In unconventional rocks, only the "*edges of the plots*" delimited by the SRV are producible, the central part cannot be recovered. This explains both the rapid decline in production and the low recovery rate (IHS CERA).

What roles does the fracturing fluid play and what does it contain?

The main function of the fracturing fluid is twofold: to fracture the rock, but also to coat the fractures with sand to support them. The fracturing fluid is therefore a water (90%) and sand (9%) suspension injected at a very high flow rate using high pressure pumps. To carry the sand to the fracture without it settling in the well, the viscosity of the water must be increased by jellifying it using chemical additives. Nonetheless, although a viscous liquid makes it easier to carry sand, it creates only wide fractures and does not stimulate the network of micro-fissures. To achieve this, water that is almost pure (i.e. no added thickeners) must be used, but this time the major disadvantage is obviously the difficulty in transporting the sand to its destination. The solution lies in carrying the sand at very high velocities so that it does not have time to settle. To this end slickwaters are pumped in at rates that are three to four times higher than thick ones. In practice, so that wide fractures can be opened and the micro-fissure network stimulated, the carrier-fluid injected is alternately thick or slickwater. In technical terms this is called hybrid fracturing.

The sand concentrations in the pumping phase are not always the same. The process generally begins with a viscous fluid containing no sand, to open the way, then sand is added and the concentration gradually increased. A typical hydraulic fracturing stage of 1,500 m^3 contains between 150 and 200 tons of sand.

In addition to creating fractures and carrying the sand the fracturing fluid must also protect the well from side effects such as chemical corrosion ("*rusting*") or biological corrosion (the bacteria present in the water are real steel-eaters!). The injection pressure must also be reduced as much as possible by lowering the friction between the fluid and the tubing so that the injection pump power remains acceptable. Low proportions of chemicals need to be added to the fracturing fluid so that it can achieve these different functions. Although they represent less than 1% of the total volume and most of them are used domestically, stakeholders see these chemicals as an environmental threat.

Reacting to this, the oil industry decided to disclose publicly the type and quantity of chemicals used for hydraulic fracturing. The data are published on a voluntary (or regulatory) basis on a public internet site called "*FracFocus*". The IOGP (International Oil & Gas Producer Association) also runs a website "*NGSfacts*" where oil and gas companies can publish the chemical additives used for hydraulic fracturing purposes.

Primary roles

Fracturing fluid **has two essential functions**. The first is to fracture the rock hydraulically and the second is to carry graded sand or micro-beads (ceramic, glass, bauxite). This *propping agent* will coat the fracture and keep it open once pumping stops (**Figure 1**). The fracturing fluid is therefore a water (90%) and sand (9%) suspension. The small remaining percentage comprises different chemicals, most of which are used to combat the side effects, but we will come back to this point later.

In practice, the suspension is pumped from the surface into the well at a very high rate (between $3m^3$ and $15m^3$ per minute) using very high pressure pumps. Remember that the pressure needed to fracture the rock must exceed the value of the minor horizontal thrust which, at a depth of 3,000m, can reach up to 500 times the atmospheric pressure. Such ranges of flow rates and pressures require extremely high pumping power, which commonly exceeds 10,000 horsepower.

So that the sand can be carried to the fracture to be coated, it must remain in suspension in the carrying fluid during the long journey from the surface to its final destination. As sand is denser[1] than water, if no precautions are taken it will rapidly settle either in the surface pipes or at the bottom of the well. The transportation of solid particles in a fluid is governed by Stokes' law[2] which states that a viscous liquid (e.g. oil) can easily carry solid particles for a long time, unlike a low-viscosity liquid (e.g. water). So, in order to carry the sand without it settling, the viscosity of the water must be increased by jellifying it using additives

1. Sand has a specific gravity of approximately 2.7 whereas water has a specific gravity of 1.
2. See Appendix 3.

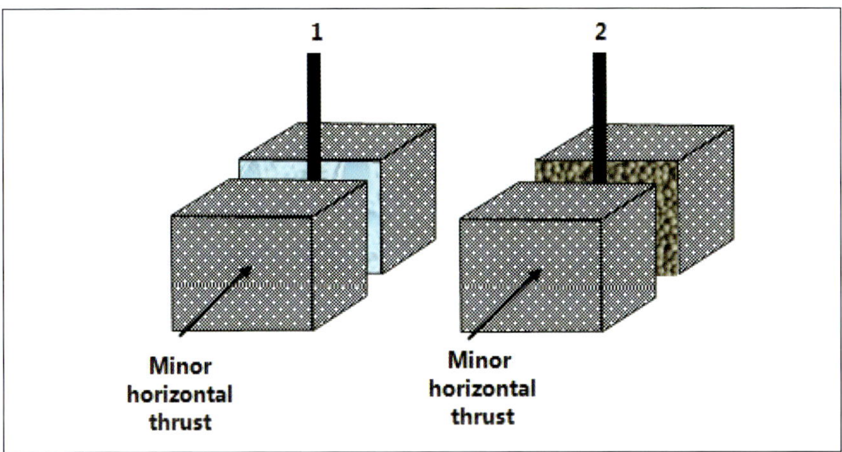

Figure 1 – The two functions of fracturing fluid. 1 – Create fractures in the rocks by increasing the fluid pressure to exceed the minor horizontal thrust. 2 – Carry sand to the fractures to keep them open and enable the gas or oil to flow to the well.

such as guar gum[3], commonly used in the food industry to thicken ice-cream or in cosmetics to manufacture lipstick.

Though a viscous fracturing fluid can easily carry sand, it does have a major disadvantage. It creates wide fractures (major highways in our road network analogy - **Figure 2**) but does not stimulate the micro-fissure network (i.e. the secondary roads network or even the little country roads) which is crucial to exploiting the smallest plots. To achieve this, pure water (i.e. no added thickeners) must be used, but here the disadvantage is the difficulty of transporting the sand to its final destination. To offset this obstacle the sand must be carried at high velocity so that it does not have time to settle. So slickwater[4] is pumped in at flow rates that are three to four times higher than those of thick liquids.

3. Guar gum is a vegetable powder derived from the guar bean, produced mainly in India. In addition to its usage in the cosmetic industry, guar gum makes bread dough soft, gives firmness to ice-cream and structure to hair conditioner. http://www.lemonde.fr/economie/article/2012/11/02/la-belle-histoire-du-guar-haricot-indien-sacre-roi-de-l-ice-cream-et-du-gaz-de-schiste_1784800_3234.html

4. Designates low-viscosity fluids as opposed to "*thick*", i.e. viscous ones.

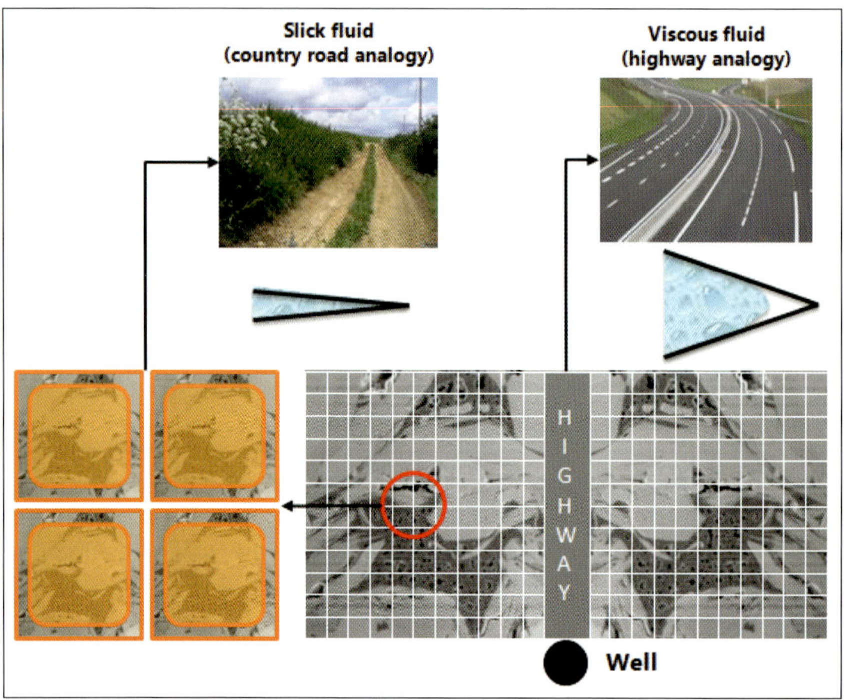

Figure 2 – A viscous fluid easily transports sand to open wide fractures but does not stimulate the micro-fissures. However, a low-viscosity liquid (pure water) penetrates micro-fissures but it has a low sand-transport capacity The low viscosity is therefore offset by a very high flow rate.

In practice, as we want to create highways, secondary roads and country roads at the same time, the thick fluids and slick fluids are injected alternately. In technical terms this is called hybrid fracturing.

Finally, average sand concentrations lie somewhere between 100 and 200 grams per liter of water but are not constant. Fracturing often begins with a pad of viscous fluid to open a wide fracture, followed by low concentrations of around 50 grams per liter of water and, in the final stages, the concentration is raised to around 400 grams or more per liter of water. A typical hydraulic fracturing stage of 1,500 m^3 contains between 150 and 200 tons of sand.

Secondary roles

In addition to its two main functions (fracturing rock and carrying sand) the fracturing fluid must also protect the well from side effects such as chemical corrosion (rusting) or biological corrosion (the bacteria present in the water are real steel-eaters!). The injection pressure must also be reduced as far as possible by lowering the friction between the fluid and the tubing so that the injection pump power remains acceptable. Very low proportions of chemicals need to be added to the fracturing fluid so that it can achieve these different functions. Although they represent less than 1% of the total volume and most of them are used domestically (agro-food, cosmetics, detergents, disinfectants - **Figure 3**), stakeholders (authorities, local communities, NGO) see these chemicals as a significant environmental risk. The oil industry is also accused of a lack of transparency by refusing, on the pretext of industrial secrecy, to disclose the exact type and quantity of chemicals used.

Transparency: the fracfocus and ngsfacts websites

The oil industry therefore decided to disclose publicly the type and quantity of chemicals used, well by well, for hydraulic fracturing. The data are published on a voluntary (or regulatory) basis on a public register called "*FracFocus*"[5] and the site is managed by two American environmental protection bodies. The aim is to provide the general public with factual transparent information and to remain neutral on the hydraulic fracturing question. The IOGP (International Oil & Gas Producer Association) also runs a website[6] where oil and gas companies can voluntarily disclose the chemical additives used for hydraulic fracturing in each individual well.

5. http://fracfocus.org
6. http://www.ngsfacts.org/

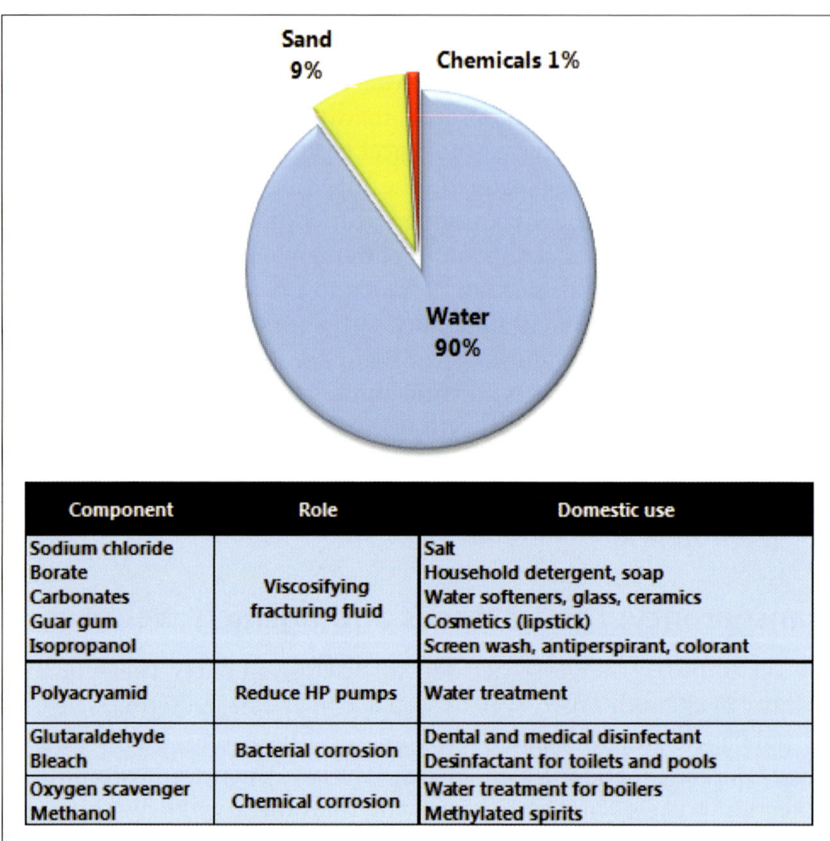

Component	Role	Domestic use
Sodium chloride Borate Carbonates Guar gum Isopropanol	Viscosifying fracturing fluid	Salt Household detergent, soap Water softeners, glass, ceramics Cosmetics (lipstick) Screen wash, antiperspirant, colorant
Polyacryamid	Reduce HP pumps	Water treatment
Glutaraldehyde Bleach	Bacterial corrosion	Dental and medical disinfectant Desinfactant for toilets and pools
Oxygen scavenger Methanol	Chemical corrosion	Water treatment for boilers Methylated spirits

Figure 3 – Typical composition of a fracturing fluid. The chemical additives represent just 1% of the total volume.
(http://fracfocus.org/chemical-use/what-chemicals-are-used)

How does the multi-fracturing process apply to a horizontal well and what equipment is used?

Hydraulic fracturing requires specific surface installations and equipment in the horizontal section of the well.

The surface installations are designed to prepare the fracturing fluid and then carry it at high pressure to inject it into the formation at a high flow rate. First, the water is mixed with the chemicals and then the sand is introduced using "*mixing augers*" very similar to those used in agriculture. The second phase consists in raising the water + sand + chemicals mixture to a high pressure, usually 500 times the atmospheric pressure. These pressure levels require pumping power equivalent to that of hundred 100Hp cars.

In the borehole, the most common method is the plug and perf, which consists in perforating the casing and the cement sheath with small explosive charges, then inserting a rubber packer between the last stage fractured and the newly-perforated zone to trap the fluid there. The operation is repeated until the start of the horizontal section is reached.

However, in the well the production from the fractured network (SRV!) converges just as traffic converges on a road and highway network on the way into a big city. Even though this network is extremely efficient, an awkward, narrow junction creates a bottleneck, the borehole equivalent of a traffic jam. The quality of the junction is therefore just as important as the network itself.

To improve this connection the first solution is to perforate the casing and the cement using a fluid under very high pressure rather than explosives. The *hydrojet* creates a clean perforation and very little collateral damage.

The second consists in keeping the rock bare (this is known as an "*open*" hole) with no casing and no cement. A pre-assembly of packers is run into the well and then inflated. The different zones to be fractured are then opened using sliding sleeves to trap the fluid in the required zone. Starting from the end of the well the operation is repeated in the same way for each fracturing stage. This method, which has the major advantage of avoiding bottlenecks by creating a direct efficient link between the well and the zone to be fractured, is applicable only if the borehole is perfectly circular. If it is not, the integrity of the packer seals would be compromised.

Hydraulic fracturing requires on the one hand dedicated surface installations, and on the other, specific equipment in the horizontal section of the well.

Surface equipment

The surface installations are designed to prepare the fracturing fluid (phase 1) and then to raise it to a high pressure (phase 2) before injecting it into the well at a high flow rate (**Figure 1**). The mixture is made in two steps. First the water (brought to the site either by truck or through a pipeline) is mixed with the chemicals, in particular those designed to thicken the fluid so that it can carry the sand. They are brought to the site in special containers and measured out specifically from a specialized unit. Next the sand is added to the water and chemical mix using augers (**Figure 1**) very similar to those commonly used in agriculture. Their rotation speed can be changed to adjust the sand concentration of the final mixture accurately.

The second phase consists in raising the water + sand + chemicals mixture to a high pressure so that it can be injected into the well at a high flow rate (between $3m^3$ and $15m^3$ per minute). Remember that to fracture the rock, the fluid pressure must exceed the value of the minor horizontal thrust which at a depth of 3,000m often reaches 500 times the atmospheric pressure. This is achieved using very high pressure pumps at the surface which usually run on diesel and are placed on trucks lined up on site. These ranges of flow rate and pressure require massive levels of pumping power, which readily exceed 10,000 horsepower, the equivalent of 100 to 150 cars (**Figure 2**).

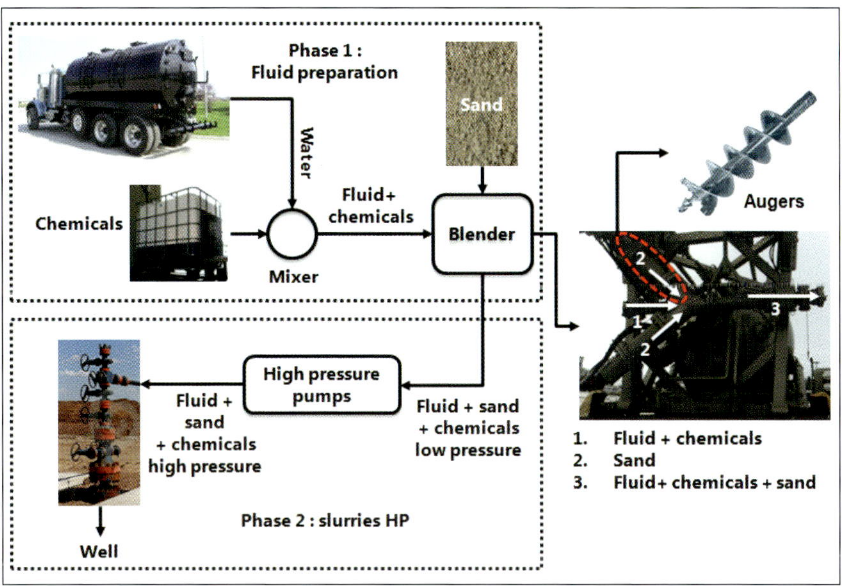

Figure 1 – Simplified plan of surface installations. Augers similar to those used in agriculture are used to adjust the concentration of sand to be mixed into the fluid which, as already mentioned, comprises almost 90% water.

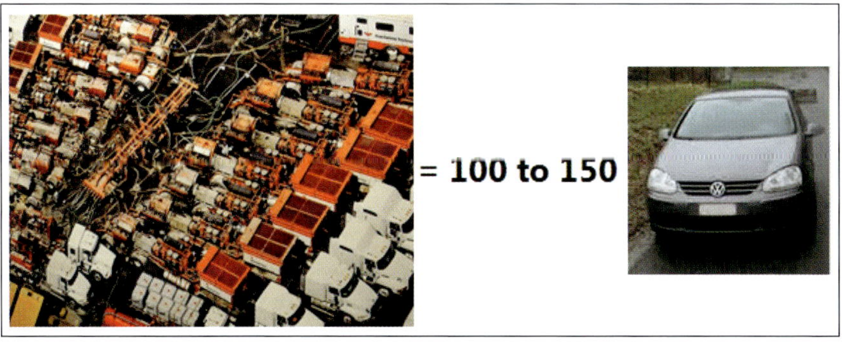

Figure 2 – The pumping power required for hydraulic fracturing is equivalent to that of 100 – 150 100hp cars.

Well equipment

In the borehole the most common method is the plug and perf. Once the horizontal part of the well has been protected by cemented casing, the fracturing stages are created starting from the end of the borehole and working back towards the opening. A well in the process of being fractured can therefore be divided into three sections (**Figure 3**): the section already fractured at the end of the well, a central zone in the process of being fractured and a zone that has not yet been fractured. Let's now have a closer look at the central zone.

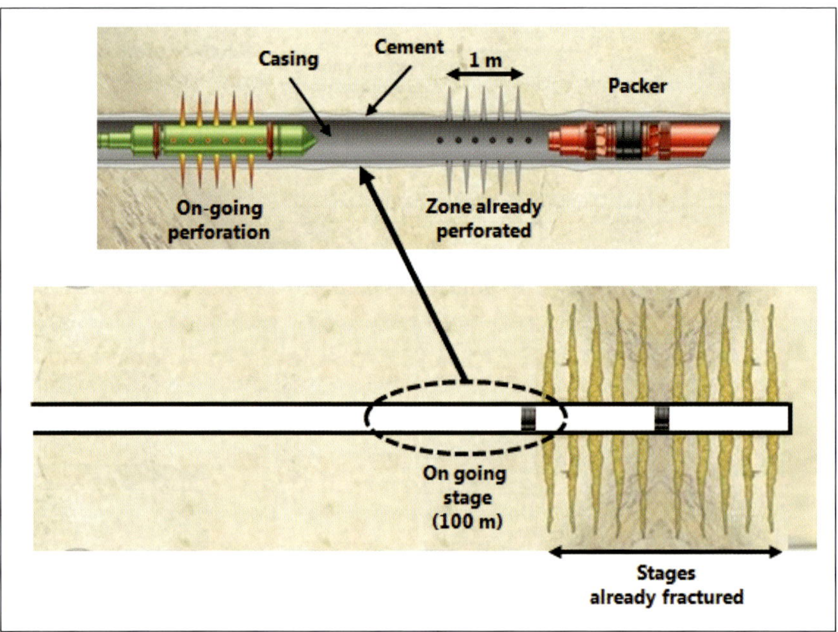

Figure 3 – Plug and perf is the most commonly used method. It consists in perforating a zone using explosive charges and then fracturing it after isolating it using a packer made of elastomer and composite material. The horizontal section is fractured from the end of the borehole working back towards the entrance. The packers are then drilled and the well brought on stream.

To create a channel between the well and the formation, perforation guns are run into the well to perforate the casing and the cement sheath over a distance of 1-2m using small explosive charges. The operation is repeated several times to cover the entire zone to be fractured. Once this has been done, a packer (a sort of watertight rubber bung) is inserted between the stage that has already been fractured and the zone that has just been perforated so that the fluid is trapped there and cannot run downstream (as it is blocked by the packer), or upstream (as the zone has not been perforated). The plug and perf operation is then repeated working back toward the inlet of the horizontal section of the borehole. The packers, which are made of composite material covered with rubber, cannot be removed and must be drilled later on to open all the perforations (and therefore the fractures) for production.

The crucial importance of the connection between the rock and the well[1]

In the well, the production from the fractured network (remember our famous SRV!) converges just as traffic converges on a road and highway network on the way into a big city. Even though the highway network is extremely efficient, a narrow junction where the tarmac is damaged creates a bottle neck equivalent to a traffic jam (**Figure 4**). The quality of the junction is therefore just as important as the network itself and the connection between the well and the SRV is no exception. Even though the quality of the SRV is exceptional, a narrow sinuous and degraded reservoir/borehole connection (fractures not open wide enough) slows the fluid down and heavily penalizes production. The care taken in creating the connection is therefore of the utmost importance.

Two ways of improving the reservoir-borehole interface

The standard perforation method involves the use of explosive charges and, although these mini-explosions perforate the steel casing and cement to create a direct connection between the formation and the

1. Also called the *"reservoir-borehole interface"*.

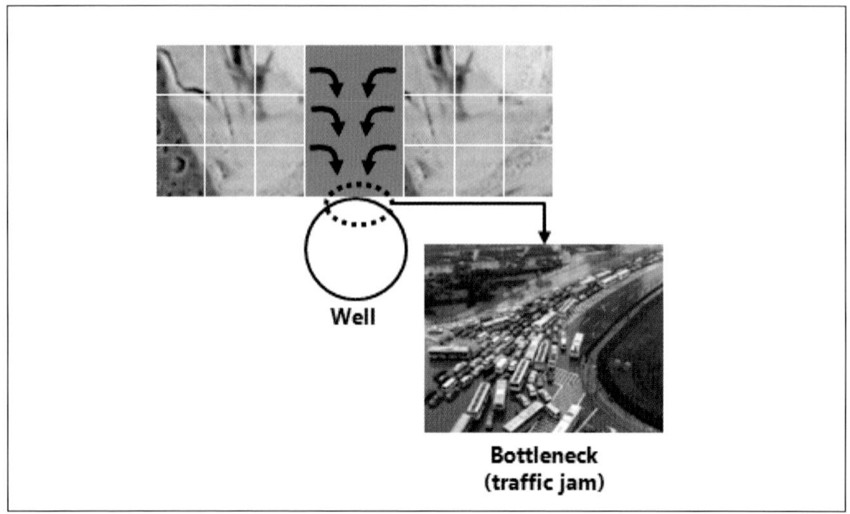

Figure 4 – The SRV production converges toward the borehole, which is a potential bottleneck zone.

borehole, they generate a compaction zone in which the rock suffers significant damage.

The first alternative method is the *hydrojet* which consists in perforating the casing and the cement using a fluid under very high pressure. Unlike explosives, the hydrojet creates a clean perforation and very little collateral damage. Even more sophisticated techniques such as lasers could revolutionize perforation technologies in the next few years.

The second alternative consists in not covering the horizontal part of the well by a casing and a cement sheath. This is called an open hole. Use of this method was widespread to develop the Bakken shale oil reservoir in North Dakota.

If, as in a cased well, a single packer was used, the entire open upstream part of the formation would be pressurized indiscriminately and it would fracture at its most fragile point. To subsequently fracture different target zones starting from the end of the borehole, the portion to be fractured must be isolated between two packers so that the fluid is confined and cannot penetrate either upstream or downstream. To do so, a pre-assembly

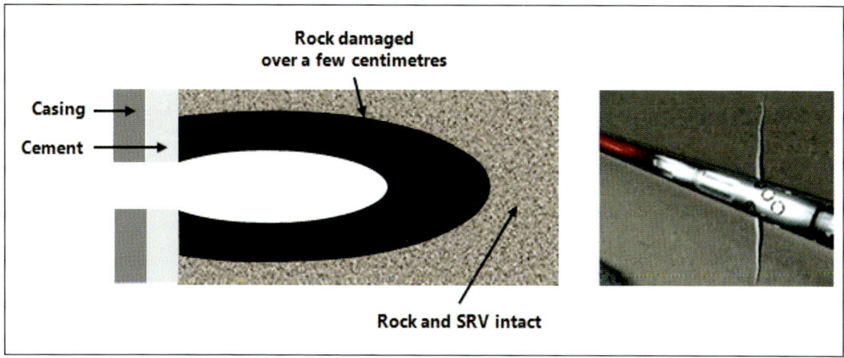

Figure 5 – Left: standard perforation. The explosive charge creates a compaction zone in which the rock suffers considerable damage. Right: the hydrojet technique creates a reservoir/borehole interface with minimal damage.

of packers[2] is run into the horizontal section of the well and then inflated. Then starting from the end of the well, the sliding sleeves located between each pair of packers are opened one by one using steel balls of increasing caliber which are injected into the well. (**Figure 6**). Once the sleeve has moved it opens access to the annular space located between two packers so that it can be selectively pressurized and then fractured. The operation is repeated in exactly the same way for each fracturing stage. Once all of the stages have been fractured, the balls are flown back and the well unblocked in the usual way.

The major advantage of this method is that it avoids the bottleneck phenomenon by creating a direct and efficient link between the well and the zone to be fractured. It also improves the operational efficiency because time is saved by not covering and cementing the horizontal section of the borehole to fire the perforations. The pre-assembled packers are run in and inflated together, which means that all the fracturing operations can be done much more quickly. However, it is applicable only if the borehole is **perfectly circular**, otherwise packer seal faults would soon become evident. Boreholes that are more oval or unstable[3] can be found in particular in regions where the tectonic thrust is intense. Beyond

2. The distances between packers and therefore the zones to be fractured are predetermined. It is possible not to fracture one of the predetermined zones but impossible to add one on.

3. See appendix 8.

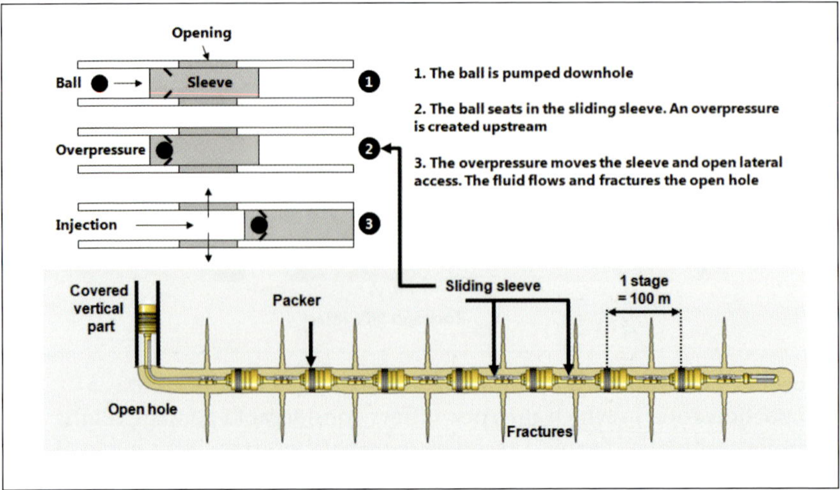

Figure 6 – Multi-fracturing in an open hole. Contrary to the plug & perf, the series of packers is run in before the operation starts. The stages are opened one after the other from the end of the well back toward the inlet using sliding sleeves that are activated using steel balls.

the fracturing operation in itself, the degradation of an open well over time may prove to be a significant constraint in the medium to long-term.

Are there any alternatives to hydraulic fracturing?

There are three possible ways of fracturing a rock, using explosion, temperature or pressure.

Used up until the end of the 1940s to stimulate oil or gas wells, detonating explosive substances (TNT and nitroglycerine) produced disappointing results and was subsequently abandoned. Deflagrating explosive substances (also called propellants) are more appealing but given their radius of impact and their inability to carry a propping agent, they are not a viable substitute for hydraulic fracturing. Deflagration could however be used as a pre-fracturing treatment to improve the interface between the borehole and the formation.

If a warm rock comes into contact with a cold fluid it cracks, but, as with explosive substances, it is not enough to fracture the rock over large enough distances and leaves no possibility for carrying a propping agent.

If pressure is the only choice to trigger fissures and transport proppants, water used as a fracturing fluid has a number of disadvantages: conflicts in terms of usage, the addition of chemicals, high capacity to damage the rock and block off part of the resources. To eliminate the need for chemicals and improve the recovery of resources, other fluids can be used.

CO_2 is a first option. Injected into the well in a cryogenic, liquid state, it is rapidly subjected to supercritical conditions when it comes into contact with the rock. It then becomes a solvent which enhances recovery. No chemicals are required, it does not damage the rock and, if injected into the subsurface layers, it contributes to limiting greenhouse gas emissions. However, this solution is no longer used today because of its poor transport capacity and implementation cost.

The main alternative to water fracturing is propane stimulation. Abundant, and easy to transport and store, this natural fluid does not damage the rock. It is more expensive than water, but as over 90% of the fluid injected is recovered, it can be recycled. The main disadvantage is its flammability. With half the specific gravity, and with a viscosity eight times lower, than that of water, it cannot carry sand without being jellified using chemical additives. Since its launch in 2008, 2,500 on- site tests have been run, essentially in Canada. The two main drawbacks of propane gel stimulation are the use of chemical products and its inability to stimulate a complex fissure network.

Using pure propane rather than propane gel is an attractive alternative but, as with CO_2, pure propane is not able to carry sand as it has a very low specific gravity and viscosity. To square the circle, these shortcomings are offset by replacing heavy sand with light glass or ceramic microbeads that can be carried in the propane. Pure propane stimulation coupled with a light propping agent would seem like a very strong candidate for replacing hydraulic fracturing in the long term. No need for water or chemicals, no recovered fluid to be treated, complex fracture networks undamaged, thereby significantly boosting reserves; a win-win technological leap forward both in terms of the environment and in terms of the sustainable production of shale oil and gas resources worldwide. Although this method is particularly suitable for arid and arctic regions, the safety standards regarding product flammability can make it difficult to implement in densely-populated urban regions.

Though there are a number of potential alternatives to hydraulic fracturing, there is no alternative to fracturing itself, since it is only by artificially fissuring the source rock to give it sufficient permeability that oil and gas can be extracted from it cost-effectively. It is therefore among the different rock fissuring methods that alternative solutions must be found. Three methods can be used to fissure a rock material[1]: dynamic fracturing using explosives, thermal fracturing and hydraulic fracturing.

Dynamic fracturing using explosives

An explosion is characterized by the extremely rapid, related movement of a front of burning gas (called the flame front) and cold gas (called the shock wave). There are two main types of explosions: **deflagration**, a

1. Luca Gandossi (2013) "An overview of alternative technologies to hydraulic fracturing for shale gas production" European Commission. JRC Scientific and Policy Report.

sort of soft explosion in which the burning gas moves more slowly than the shock wave and **detonation** which is much more violent as the flame front and the shock wave occur simultaneously.

A detonation increases pressure significantly and rapidly, displaces a huge volume of air and generates a tremendous noise. The use of the detonating power of an explosive to fissure a rock is the most common mining excavation technique and was also frequently used up until the end of the 1940s to stimulate the production of oil or gas-bearing formations. It consisted in exploding liquid nitroglycerine or TNT pastilles[2] at the bottom of a well. At the end of the 1960s, the Soviet Union had even experimented nuclear energy to fracture rock formations but, in a confined environment like the bottom of a well, the abruptness of the impulse (very high pressure reached in just a few micro seconds) compacts the rock rather than fragmenting it, thereby reducing the permeability around the borehole. The disappointing and unpredictable results led to this technique being abandoned in favor of hydraulic fracturing.

Unlike detonating explosive substances, deflagrating substances (also called propellants) offer a possible alternative to fracturing in the production of shale oil and gas. When they deflagrate, propellants rapidly generate gases under very high pressure (1,500 times the atmospheric pressure is reached in 10 milliseconds). However, even though this is ten thousand times faster than for hydraulic fracturing, it is ten thousand times slower than for detonating explosive substances. Moreover, unlike detonation, deflagration does not generate a shock wave, so the rock is fissured rather than being violently compacted.

Although one of the advantages of this technique is that it requires neither water nor chemicals to create complex fracture networks, the penetration of the latter is restricted to a few meters around the borehole (compared with several dozens of meters for hydraulic fracturing). In addition, as the propellant cannot carry a propping agent into the fractures, their conductivity relies on the hit-and-miss hypothesis that they will remain open naturally after undergoing a micro-sliding effect. As a final point, although relatively low levels of energy are released during the deflagration, it is intense enough to trigger seismic events. Even if the use

2. Trinitrotoluene (TNT) is an explosive used in many mixtures, in particular in equal proportions with ammonium nitrate to form Amatol.

Advantages	Disadvantages
No water, no chemicals	Very limited penetration
Complex fracture networks	Do not transport proppant
Low cost, light equipment	Can induce seismic events

Figure 1 – Deflagrating explosive substances. Advantages and disadvantages compared with hydraulic fracturing.

of propellants cannot replace hydraulic fracturing under any circumstances, it could nonetheless be applied as a pre-fracturing treatment to improve the connection between the well and the SRV[3].

Cryogenic fracturing

Who as a child has never tried heating a glass ball and then plunging it into cold water? When it comes into contact with the cold water the glass ball, subjected to what are called "*thermal constraints*", cracks. Similarly, a warm rock (remember that the average geothermal gradient is, 30°C/1000 m) that comes into contact with a cold fluid, will crack.

A particularly popular cryogenic compound[4] is carbon dioxide (CO_2) which in its liquid form is used as a refrigerant to freeze certain foodstuffs, or as a fire extinguishing agent[5]. As shown in the phase diagram **Figure 2**, carbon dioxide liquefies at -56.4°C and at just 5 times the atmospheric pressure, but remains a liquid refrigerant at pressures of between 100 and 1,000 times the atmospheric pressure, the range of values for bottomhole pressure.

However, unless the rock cools down over a very long period of time, the contact of a cold fluid with a warm rock is not enough to fracture the rock over long enough distances. In fact, the proposed technique[6] aims

3. See question 9 "How does the multi-fracturing process apply to a horizontal well and what equipment is used?
4. In Greek, "*cryogenic*" means "*the production of freezing cold*".
5. This is often referred to as "*dry ice*," as the liquid CO_2 solidifies immediately as it is pumped out of the extinguisher, producing a white powder.
6. Mueller M., Amro M., *et al.* (2012) "Stimulation of Tight Gas Reservoir using coupled Hydraulic and CO_2 Cold-frac Technology", SPE Asia Pacific Oil and Gas Conference and Exhibition. Perth, Australia, Society of Petroleum Engineers.

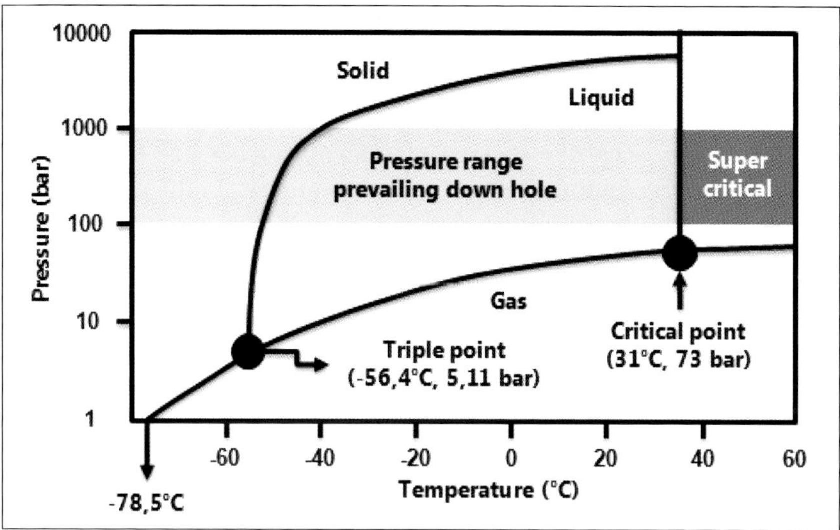

Figure 2 – Carbon dioxide phase diagram. In the bottomhole pressure range, the CO_2 is first a liquid; then, as it heats up, it changes into the supercritical state and becomes a solvent.

not to facture the rock using cooling alone, but to enhance standard hydraulic fracturing techniques using the thermal effect by injecting large quantities of liquid CO_2 under high pressure. Depending on the injection rate, the temperature can drop to between 50°C and 100°C.

Alternative stimulation fluids

The processes of dynamic and cryogenic fracturing which are briefly described above, do not alone suffice to fracture the rock over long enough distances or to carry the propping agent. Even though these methods could be used as possible catalysts either to improve the well/borehole interface (deflagration) or to enhance the efficiency of the process (cryogenic fluid), the only method able to create and maintain a fracture network is the high pressure injection of a proppant-carrying liquid. However, even though there is currently no real substitute for hydraulic fracturing, there are other alternatives to using water as the fracturing fluid.

Although cheap and usually available in large quantities (except in arid zones) water also has numerous drawbacks. Apart from arguments about the use of water[7], chemical additives[8] need to be used to counteract chemical and bacterial corrosion. But the main disadvantage, often ignored by the general public, is its ability to damage the source rock (**Figure 3**) and consequently to jeopardize production and write off a significant percentage of recoverable resources. The damage is manifested first and foremost by the creation of retention forces[9] in the microfissures in the rock. The proof is that only 20-40% of the volume injected is recovered when the well is cleaned up. The oil and gas located upstream of the trapped water will never be recovered. Source rocks also contain clay minerals which swell as they adsorb water and block up many

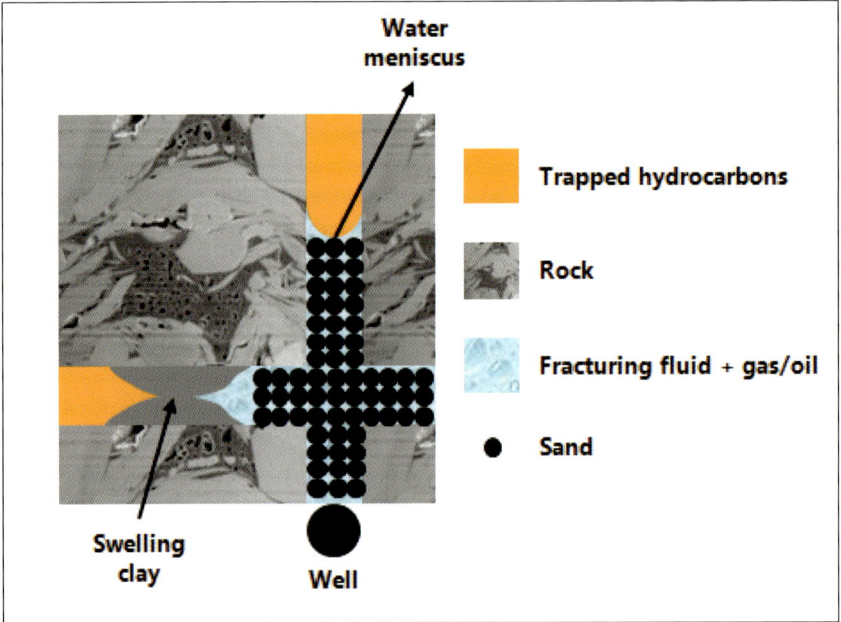

Figure 3 – Water creates two types of damage. Its high surface tension creates small menisci which block many of the microfissures. Moreover, certain clay minerals can swell in contact with water and plug other microfissures. The oil and gas located upstream of these barriers are permanently trapped. The combination of physical damage (menisci) and chemical damage (swelling) lead to a significant loss of production and recoverable resources.

7. The issue of usage conflicts is examined in question 14.
8. See question 8.
9. The technical term is *"capillary forces"*.

microfissures, causing additional production losses and the loss of ultimate resources. The damage is so crippling that it is estimated that in certain cases, less than 20% of the fissures created by the fracturing process actually contribute to production. Improving the reserves recovery therefore requires the use of a less damaging fluid. Among the different alternatives, liquid CO_2 (mentioned in the previous paragraph for its cryogenic properties) and LPG (Liquid Petroleum Gas – essentially propane) are the two most attractive options.

In addition to its cooling capacity, CO_2 has remarkable physical properties. When it is subjected to supercritical conditions (over 31°C and 73 bars - **Figure 2**), it becomes a solvent[10] and when it mixes with the oil and gas, improves the recovery rate. When injected into the rock in a liquid cryogenic state, CO_2 rapidly reaches supercritical conditions as it heats up in contact with the rock. When the well is put on stream, the fluid depressurizes and becomes a gas which, compared with water, significantly improves the efficiency of the clean up process. Unlike water, CO_2 does not damage the rock or create capillary forces, or cause clay minerals to swell. It requires only a small quantity of chemical additives (**Figure 4**). A final advantage is that it can be adsorbed by shales. Fracturing using CO_2 could be an advantageous underground storage method and contribute to limiting greenhouse gas emissions. However, in view of its low viscosity, the transport capacity of CO_2 is poor, and the concentration of propping agents is therefore very limited. CO_2 fracturing was widely-used in the production of conventional resources at the beginning of the 2000s, but today this method has been practically abandoned.

The main and most popular alternative to water today is stimulation using jellified LPG, a base fluid comprising 96% propane (C_3H_8), a natural fuel commonly used for domestic purposes and for which the transport and storage standards are clearly defined in most countries in the world. Unlike CO_2 it is not a cryogenic fluid; it is transported and stored at a low pressure (15 bars) and ambient temperature. It is available in sufficient quantities in many oil-producing countries as a by-product of crude oil. Like CO_2, LPG has many advantages over water. It does not damage the rock (no capillary menisci or clay mineral swelling) and does not corrode

10. This property is used to extract caffeine from coffee to obtain decaffeinated coffee.

Physical properties	Water	LPG	CO_2
Viscosity (cP)	0.66	0,08	0,1
Density	1,02	0,51	0,5
Clay sensitivity	☹	☺	☺
Corrosion	☹	☺	☹

Function	CO_2 liquid	LPG
Avail & storage	☹	😐
Water & chemicals	☺	☺
Sand transport	☹	😐
Clean up	☺	☺
Damage	☺	☺
Cost	☹	☹
Safety	☺	☹

Figure 4 – LPG (essentially propane gel) and liquid CO_2 have physical properties which give them considerable advantages over water. In particular, they do not damage the rock, require little or no chemical additives, and make the well clean up easier thanks to their low specific gravity. They are more expensive but, as 90% of the fluid is recovered, the method remains economically viable.

steel (no need for anti-corrosion chemicals). Its low specific gravity (half that of water) makes the well clean up easier. Although it is much more expensive than water (between US$ 350 and US$ 500 per cubic meter), it can easily be recycled as over 90% of the fluid injected is recovered[11].

As propane has a specific gravity which is two times lower than that of water, and a viscosity eight times lower, it cannot carry sand (specific gravity of 2.7) in its normal state, so it is jellified using a number of chemical thickeners before being injected. Its main disadvantage is its flammability, and using it requires many precautionary measures (safety perimeter, a series of control measures) to ensure the safety of personnel working on site.

Since its industrial launch in 2008, the Canadian company GASFRAC[12] has carried out over 2,500 tests mainly in Canada and the United States. The results of the production of two wells, one fractured using water and the other stimulated using LPG are compared in **Figure 6**. In spite of an

11. The 90-95% recovery confirms the absence of capillary forces.

12. http://www.gasfrac.com/

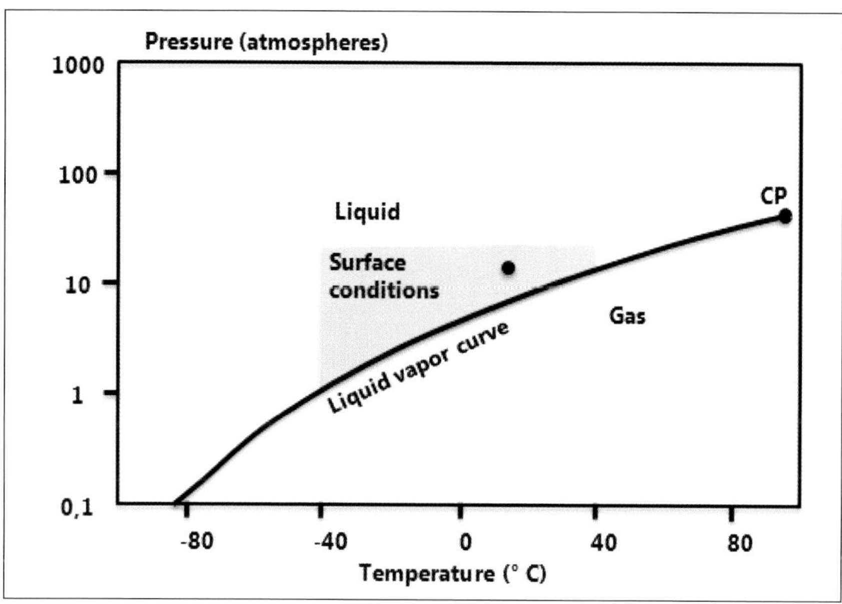

Figure 5 – Propane phase diagram. It remains a liquid during the fracturing operation, then most of it vaporizes in contact with the gas. It is soluble in crude oil and improves its market quality.

injected volume that is 75% lower, the well stimulated using LPG produced twice as much as the well fractured using water at the end of an eight-month period. According to oil and gas prices, LPG stimulation could therefore prove cost-effective as long as the total product volume can be recycled.

Jellified propane stimulation has two main disadvantages. On the one hand, the jellification of the fluid requires the use of chemicals which are not usually environmentally friendly. On the other hand, regarding production, jellified propane creates wide, planar fractures (highways!) and not the networks of complex fissures (the small country roads!) crucial to the production of sufficiently small plots. Using pure, rather than jellified, propane would therefore seem like a very attractive option: no need for chemical additives, and a low-viscosity fluid that stimulates all the microfissures without damaging them. However, this is somewhat a pipe dream as sand, with its specific gravity of 2.7, cannot be carried by a low viscosity fluid with a specific gravity of less than 0.5.

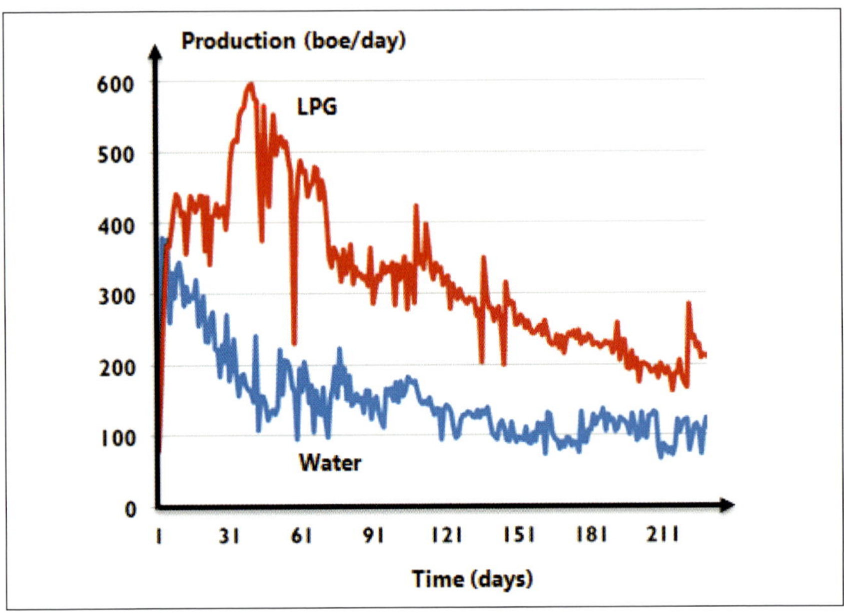

Figure 6 – Comparison between two production graphs. After a period of 8 months, the well stimulated using LPG has produced double the quantity of the well fractured using water and for an injected volume 75% lower (source: Black Brusk O&G PL).

The American company EcorpStim[13] plans to square the circle by compensating for the lack of specific gravity and fluid viscosity by concentrating on the specific gravity and size of the particles that make up the propping agent[14]. So, by replacing heavy coarse grains of sand by light glass or ceramic[15] microbeads, it is perfectly feasible to transport the propping agent to the microfracture network in the propane. Pure propane stimulation together with a light propping agent is an attractive alternative to replace standard hydraulic fracturing in the long term: no need for water or chemicals, no recovered fluid to be treated, intact complex fracture networks that significantly boost ultimate resources (**Figure 7**). Propane stimulation is a win-win technological leap forward in environmental terms, but also in terms of the sustainable development

13. http://www.ecorpstim.com

14. Remember Stokes' law. Question 8 and appendix 3.

15. http://www.oxanematerials.com/

of shale oils and gases worldwide. The method is particularly attractive in arid regions where water availability is problematic, and also in arctic and subarctic areas where, unlike water, propane remains a liquid even at very low temperatures. The safety standards governing flammability (in particular the safety perimeters) can sometimes make use of this method difficult in densely-populated urban areas.

Function	Jellified LPG	Pure LPG + light proppant
Chemicals	☹	☺
Transport proppant	☺	☺
Clean up	☺	☺
Damage	☺	☺
SRV complexity	☹	☺

Figure 7 – Stimulation using pure propane has many advantages over propane gel. No chemicals are required and complex SRV can be performed.

How can shale oil and gas be developed? How can the American revolution be exported?

A conventional offshore field is developed using a few dozen wells, each producing several thousand boe/day over long periods of time. Although the cost of development tops several tens of millions of US dollars, the wells are extremely profitable. In comparison, shale oil and gas wells produce only a few hundred boe/day, production declines very rapidly and thousands, or sometimes tens of thousands, of horizontal multi-fractured wells are required.

From a hand-made process in conventional, there is a switch to a factory development process. Well architectures and operating procedures are simplified and standardized, and consumables and services standardized and negotiated in large quantities to obtain wholesale prices in a competitive market. However, simplification and standardization will go hand in hand with stringent safety and environmental regulations. The standardization process is based in particular on the PAD concept which consists in drilling between five and fifteen horizontal wells from a surface area equivalent to that of two to three football pitches. So a field that requires 1,000 wells could be developed using just a few dozen PADs.

Implementation of the factory development process brought success to the United States. A well drilled there is three to five times cheaper than elsewhere. The method consists in proceeding by trial and error and drilling a great many cheap wells without any detailed analysis. It does not really matter if a significant percentage of them are not profitable because the successes pay for the failures.

The economic context and the availability of equipment outside the US are such that the method cannot be exported elsewhere as it stands. Although the rules of factory development are still valid, the number of unprofitable wells must be restricted by identifying as soon as possible target zones with a maximum potential called sweet spots. The quality of the zone is first evaluated during a pilot phase (a few dozen wells) to prepare a cost-effective development scheme.

Drilling and fracturing expenses represent between 70% and 90% of overall investments and will determine the economic legitimacy of a project. The fixed costs of drilling and fracturing weigh heaviest in the balance, hence the crucial importance of standardization and market fluidity, and any monopoly situation inevitably leads to spiraling costs.

Factory development strongly affects the way in which things are organized. The standard sequential process is substituted by a looped continuous learning process, each well becoming first an exploration well, then an appraisal well, a development well and, finally, a production well. Speed and number require the implementation of transverse fully-integrated organizations operating from remote centers in which all the pertinent competencies and responsibilities are brought together.

Finally, whereas shale oil projects are still profitable for well costs of over US$ 25 million, for gas, the profitability threshold falls on average to US$ 12.5 million per well in Europe and US$ 6 million per well in the US. With barrel prices above US$ 85/bbl the monetization of shale oils does not encounter any real economic stumbling blocks, whereas the profitability of gas projects requires considerable efforts when it comes to competitiveness. However with an oil barrel below US$ 70, the profitability of shale oil projects can be questioned.

Success in the US thanks to the "*trial and error*" method

A conventional offshore (e.g. North Sea) field is usually developed using a few dozen, or in exceptional cases, about a hundred wells. Each one will produce several thousand barrels per day over periods of between ten and twenty years or more. In the case of exceptional "*nuggets*" like certain fields in the ultra deep offshore[1], the production from certain wells can reach several tens of thousands of barrels per day. Such wells are usually vertical or deviated, sometimes horizontal, but almost never fractured. Although the cost of development routinely tops several tens of millions, or even a hundred million dollars[2], the total production[3] (approximately

1. e.g. offshore Angola.
2. As is the case with deep offshore wells drilled using dynamic positioning vessels.
3. Also called associated reserves.

15 Mboe per well) makes such developments extremely profitable. Remember that, on average, the cost of the wells[4] will represent 30% of overall investments[5] (**Figure 1**). In the case of heavy surface investments (e.g. for an LNG plant[6]) this can decrease to below 10%.

Owing to their very low permeability, shale oil and gas wells behave in a way that is almost the exact opposite to the way in which wells on standard fields behave (**Figure 1**). On average they produce only a few hundred boe/day and decline very rapidly (after one or two years, the production is only a tenth of the initial production!). The total production will reach just a few hundred thousand barrels per well. To maintain a certain overall production level, a declining well must therefore be replaced immediately by a new well, and so on and so forth. Thousands,

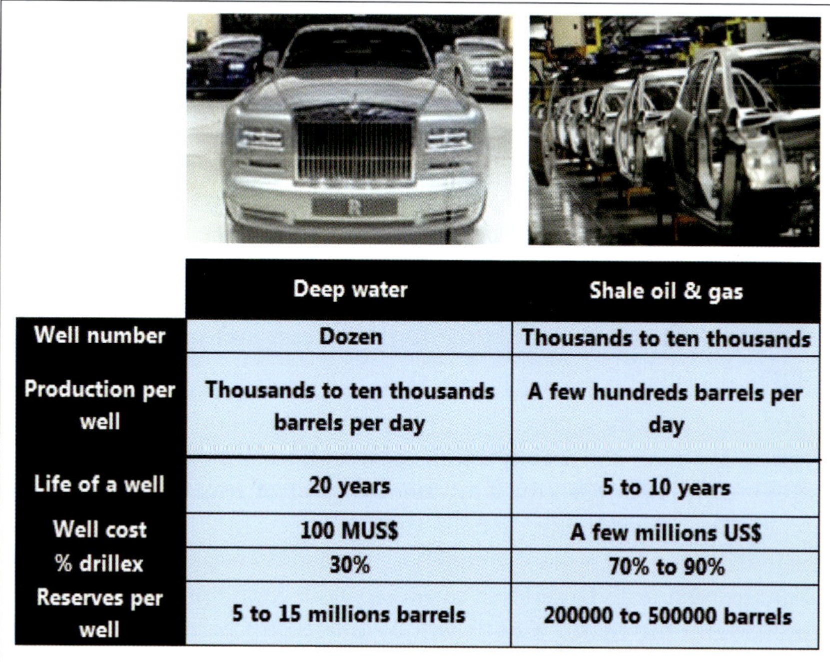

	Deep water	Shale oil & gas
Well number	Dozen	Thousands to ten thousands
Production per well	Thousands to ten thousands barrels per day	A few hundreds barrels per day
Life of a well	20 years	5 to 10 years
Well cost	100 MUS$	A few millions US$
% drillex	30%	70% to 90%
Reserves per well	5 to 15 millions barrels	200000 to 500000 barrels

Figure 1 – Comparative figures for deep water environments and shale oil and gas numbers. The development models are diametrically opposed.

4. DRILLEX =Drilling Expenditures.
5. CAPEX = Capital Expenditures.
6. LNG = Liquefied Natural Gas.

or even tens of thousands of horizontal, multi-fractured wells will be required. So, between the beginning of the American revolution and the end of 2013, over sixty thousand wells were drilled in the US. Regarding unconventional resources, the cost of the wells alone can represent 70 - 90% of investments.

The automobile industry is a very good analogy. In the case of conventional resources, each well represents a single hand-made model (a Rolls Royce) and the manufacturing costs will be of relatively little importance in that the profit margin will be extremely high. Shale oil and gas, on the other hand, must be developed with low profit margins and, in this scenario, the cost of manufacturing a well becomes the key economic element. There is a transition from a handmade process to a factory development process (**Figure 1**). Well architectures (number of casings) are simplified and standardized and the operating processes (moving drilling rigs and fracturing teams, running in casing) have to be organized so that they are run as efficiently as possible. The consumables (tubes, drilling mud, cement, sand, chemicals) and services (drilling, fracturing, borehole measurements) are standardized and negotiated in very large quantities over lengthy periods of time as part of vast contracts enabling wholesale prices to be obtained in an open, cut-throat market. However, simplification and standardization **will go hand in hand with extremely stringent safety and environmental rules** providing the conditions essential to the success and sustainability of a project.

The standardization process is among others based on the PAD concept (**Figure 2**) which consists in drilling several horizontal wells (between 5 and 20) from a surface area equivalent to that of two to three football pitches. So, for example, a field that requires 1,000 wells could be developed using just a few dozen PADS. The PADS are then connected via a main trunk line and a series of secondary collection lines to a processing center where the hydrocarbons will be treated (in particular, separated from water) so that they comply with market specifications.

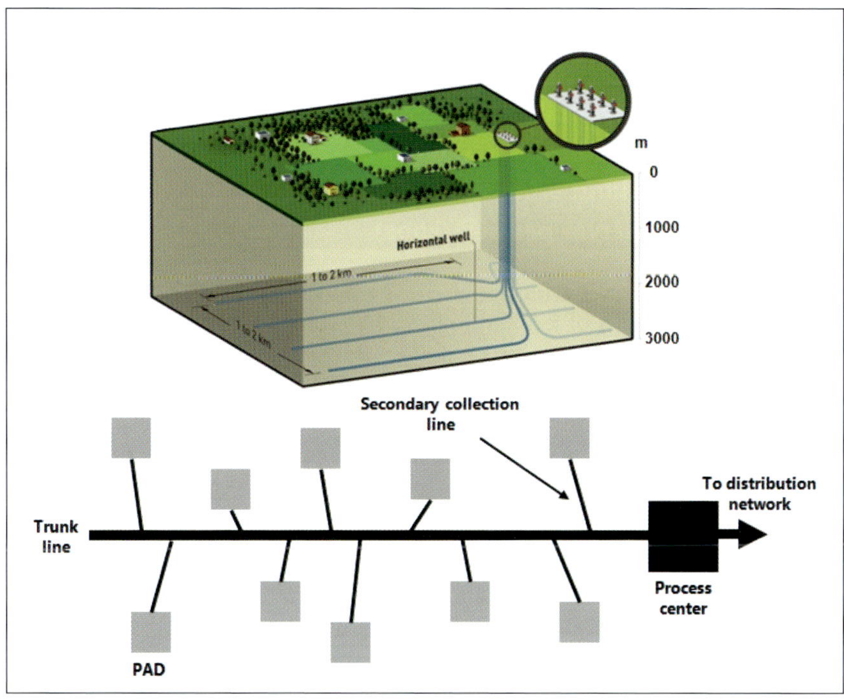

Figure 2 – Unlike the "ancestral" developments using vertical wells, the PAD concept enables several horizontal wells to be drilled from a reduced surface area. PADs are connected to the treatment center by means of a main trunk line and secondary collection lines.

Implementation of the factory development process was one of the main factors in the success of the US revolution. Of the 2,500 drilling rigs present in the world about 2,000 operate in North America. The Majors (Exxon, Shell, ConocoPhillips, Chevron, Total) and the mega service companies (Baker, Schlumberger, Halliburton, Weatherford) are challenged from all sides by independent operators and small drilling or fracturing companies in an open and highly competitive market offering unprecedented price-performance ratios. In the Barnet shale (North Texas - **Figure 3**), a well drilled to a depth of 2,500 m with a horizontal extended reach of 1,500 m with 15 fracturing stages, costs just 3 million dollars. The same well in another region of the world would be three to five times more expensive.

Drilling time (days)	12
Depth (m)	2500
Horizontal (m)	1500
Number of frac stages	15
Drilling + frac costs (MUS$)	3
Reserves per well (kboe)	500

Figure 3 – Operational performances in the Barnett shale (North Texas). A well drilled to a depth of 2,500m followed by a horizontal one 1,500m long with 15 fracturing stages for just 3 million dollars. Source: Chesapeake

The trial and error method consists in drilling and fracturing a large number of cheap wells without analyzing the geological attributes in detail, and accepting that, statistically a significant percentage of the wells will not be profitable. It doesn't really matter because the successes pay for the failures, a phenomenon for which **Figure 4** is a perfect illustration. The production map (left) is shown alongside the well location map (right), and over a third of the wells were effectively drilled in the non-productive zone.

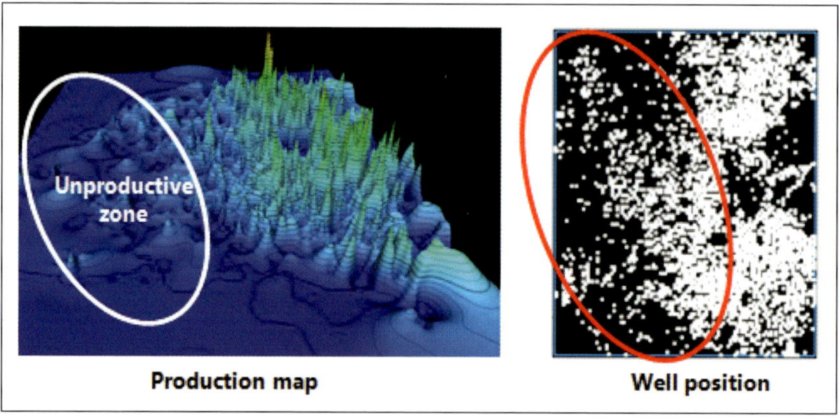

Figure 4 – The trial & error method which underpins the US development model. More than a third of the wells were drilled outside the productive zone. Source: Schlumberger Terratek

Can the US model be exported?

The economic context outside the United States (limited availability of drilling rigs and fracturing fleets, cost of services and consumables) is such that the US trial and error method cannot be exported elsewhere as it stands. Although the factory development rules (streamlined, standardized well architectures, operating processes planned as efficiently as possible, standardized consumables and services negotiated as part of long-term contracts in an open competitive market) still apply, it is crucial to minimize the number of uneconomic wells drilled outside the productive zone.

Sweet spots are areas which associate a high quality source rock (thick and rich in hydrocarbons) and an aptitude (presence of natural fractures, brittle rock) to generate complex and extended fissuring by hydraulic fracturing[7]. To identify the sweet spots as quickly as possible, during the pilot phase (see following paragraph) a number of measures are taken in the laboratory, in the wells and from seismic surveys[8] (**Figure 5**).

Pilot phase

A license (typically between 10,000 km^2 and 25,000 km^2) is much too extensive to be developed in a single phase and is therefore divided into several zones of about 2,500 km^2 each. To give an idea of size and quantity, a horizontal multi-fractured well 1,500km long drains a subsurface area of approximately 1 km^2. To develop an area of 2,500 km^2, 2,500 wells need to be drilled, fractured and connected. If we assume that each PAD comprises 10 wells, our development would therefore require 250 PADs, i.e. on average one every 3km.

However, before embarking on a massive development project, evaluating the quality of the zone to be developed is essential. This is done by drilling a certain number of test wells to identify the sweet spots and minimize the number of unproductive wells. The pilot phase can be divided into three successive steps (**Figure 6**). The first is exploratory and consists in collecting and analyzing existing data (e.g. from old abandoned wells) and then drilling several vertical wells (5 in the case of **Figure 6**).

7. See question 7.
8. See Appendix 6.

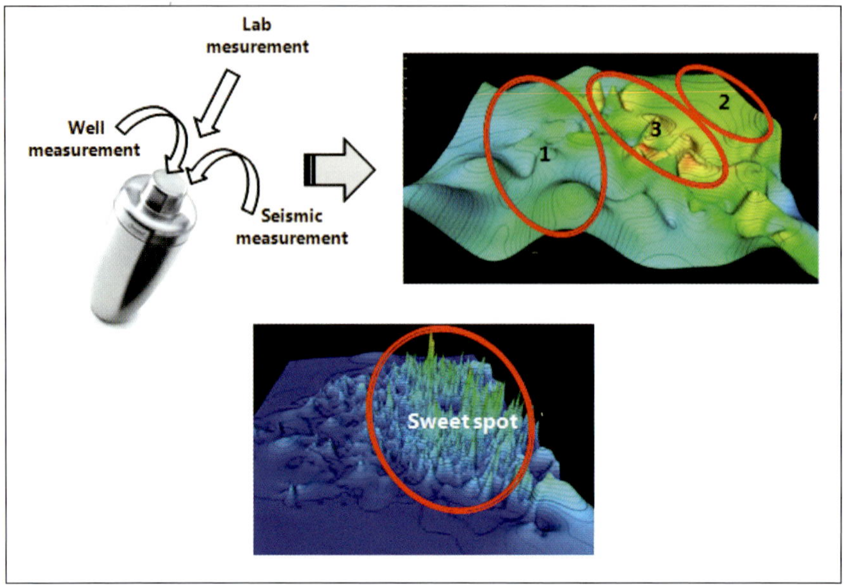

Figure 5 – Example of a sweet spot map. Three distinct areas are highlighted: Zone 1 – poor, Zone 2 – medium and zone 3 - good. Only zones 2 and 3 produce. Source: Schlumberger Terratek

Well measurements (wireline logs) and rock samples (cores or drilling cuttings from the wells) are then taken to assess the quality of the source rock. This first series of vertical wells is not likely to provide any production data. The next step is an appraisal phase which consists in drilling about ten horizontal, multi-fractured wells that will be tested (this time in production) over several months. Around the best zone (zone 5 in **Figure 6**) one or two experimental PADs will be created to estimate the commercial viability of a development. The pilot phase will last for two to three years on average. Outside the United States and according to the number of wells, it will cost between 250 and 500 million dollars. It is not so different from a conventional development in that no more than a few dozen wells will be drilled.

If the main purpose of the pilot phase is to de-risk the geological uncertainties in preparation for an economic development scheme it is also essential to evaluate and deal with acceptability issues[9] as early as possible. Although the pilot phase can also be the starting point of

9. Acceptability issues are discussed from question 12 onward

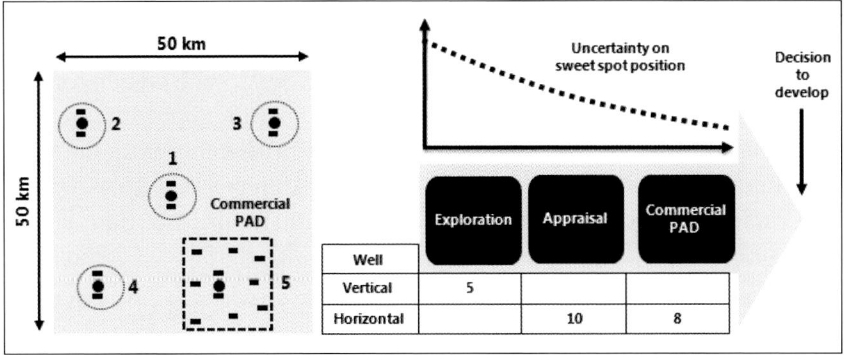

Figure 6 – The pilot phase includes and exploratory step, an appraisal step and a commercial test step.

operational knowledge (i.e. improvement of drilling and fracturing performance), the main objectives (de-risking and acceptability) must not be lost by neglecting certain crucial measures on the pretext of cost reduction.

Development phase

The positive (or negative) results of the pilot phase will determine whether or not the entire zone should be developed. It is therefore at this stage that one switches from a conventional model to a factory development model. The latter is constrained by two variables: **speed** (a hectic race to drill, fracture and connect, to replace declining wells with new ones) and **number**. During this process several tens (or even several hundreds) of PADs must be created and connected up to a treatment center from which the gas and/or oil are exported to consumers. The need to standardize equipment and procedures to obtain wholesale prices does not mean freezing the development process. By making a clear distinction between the parameters that must stay fixed (well architectures) from other modifiable parameters (number and intervals between fracturing stages), the operating model can withstand low costs and rapid cycles, with corrections being made on the job, when certain geological, technical or commercial indicators require it.

Inevitably this affects the way in which things are organized. The standard sequential process of exploration/appraisal/development/

production, during which the different disciplines (geosciences, drilling, pre-project, project, production) hand over the baton, becomes obsolete. It is replaced by a looped continuous learning process, each well becoming an exploration well, then an appraisal, development and production well (**Figure 7**). During the development phase, the geological data are continuously collected to narrow down the location of sweet spots. Owing to a rapid operational optimization, a reduction in drilling time is also expected.

Speed and number require the implementation not just of a modular, standardized approach, but also of organizations that are fully integrated and in which the different disciplines involved (geology, drilling, geomechanics, production, logistics) are answerable to a single hierarchy and their performance assessed on the basis of common shared indicators. This kind of model highlights the pertinence of open spaces and remote operational centers where all the competences and responsibilities are brought together so that the right decisions can be made collectively within the set lead-times.

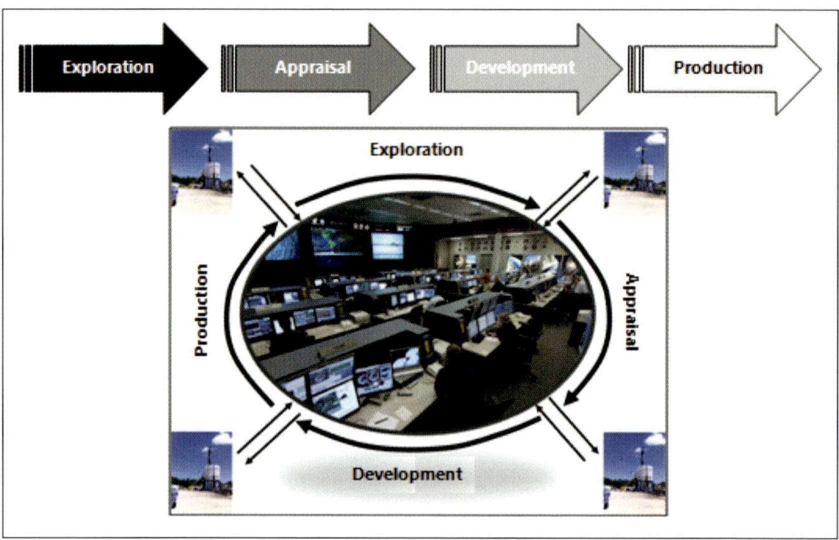

Figure 7 – Conventional phased process (top). Unconventional loop process (bottom) with an integrated operations center.

What are the economic fundamentals of a shale oil and gas project?

Drilling and fracturing expenses represent between 70% and 90% of overall investments. The parameters related to these activities therefore determine the economic legitimacy of a shale oil and gas development project. The parameters can be divided into two main categories: on the one hand, **cost-related** parameters (temporal costs such as the daily rental of the drilling rig or associated services and fixed costs such as drilling[10], fracturing[11] and collection[12] consumables) and, on the other, those that are **time-related** (drilling and fracturing time, waiting time between the finalization of a PAD and its connection). However, the income depends on the production profile (which in turn is dependent on the drilling program and the rate at which the wells are brought on stream) and the price of the gas and/or oil. Lastly, the final profit (Net Present Value and Internal Rate of Return[13]) will be a result of the comparison of expenses and income, considering the discount rate and the tax legislation in the producer country (**Figure 8**).

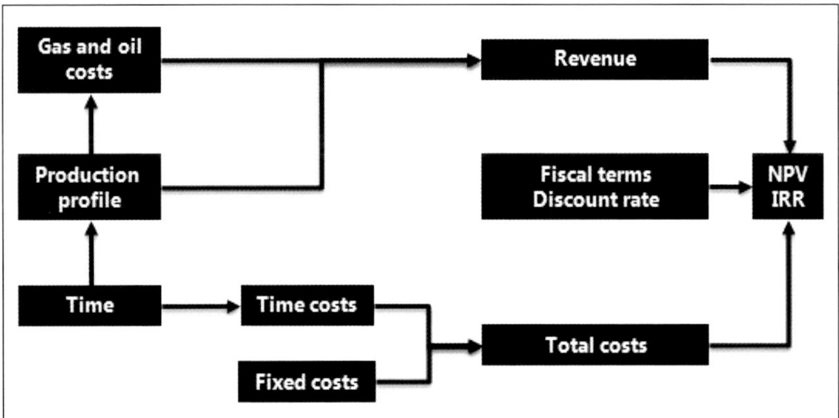

Figure 8 – The NPV and IRR of a project are calculated based on the expenses, income, discount rate and tax legislation conditions.

10. Tubes, mud, cement.
11. Water, sand, chemicals, pumps, services related to fracturing.
12. Purchase and laying of pipelines to export the gas and/or oil from the PADS to the processing center.
13. See Appendix 7.

According to their respective impact (positive or negative - **Figure 9**) on the NPV and the IRR, costs and times can be divided into three different categories:

- Fixed costs for drilling and fracturing which affect each well have a major influence on the economic viability of the project.
- Surface fixed costs (mainly the pipelines that connect the PADS to the treatment center) and the duration of the drilling program have an average impact on the economic viability of the project.
- Unless the time taken to connect the PADS is disproportionate, it has only a minor influence on the economic viability of the project.

The results reflect the crucial importance of standardizing well and surface equipment purchased in large quantities and negotiated in the context of extended contracts in a highly-competitive market. Any kind of monopoly (public or private) in the services sector (drilling or fracturing contractors) or equipment and consumables suppliers (tubing, mud, cement, sand, chemicals) will inevitably lead to spiraling costs and to the economic non-viability of the project.

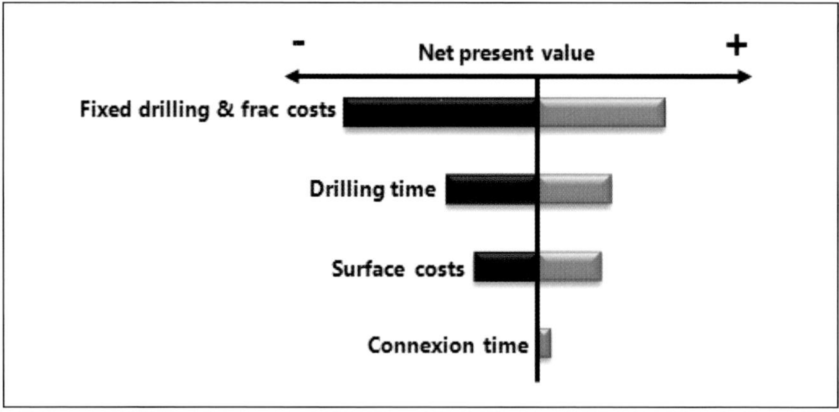

Figure 9 – Positive or negative impacts, based on a reference case, of the main costs and times on the NPV of a shale oil and gas project.

Well equilibrium cost

To highlight the difference in economic profitability between a gas project and an oil project, **Figure 10** compares two similar projects (around 500 kboe/well over a 30 years life). Whereas for oil (US$ 100/bbl), the project remains profitable for well integrated costs up to 25 MUS$, for gas (US$ 10/MBTU), the profitability treshold (considered at 15% IRR) drops to 13,5 MUS$/well.

The monetization of shale oils (e.g. the Bazhenov in Russia) do not therefore encounter any real economic obstacles above US$ 85/bbl. The profitability of gas projects (e.g. on the continent of Europe or in China) remains marginal and their set up will require considerable efforts to boost industrial competitiveness (regarding the price of equipment and services) to reduce the overall cost of the wells. However with an oil barrel below US$ 70, the profitability of shale oil projects can be questioned.

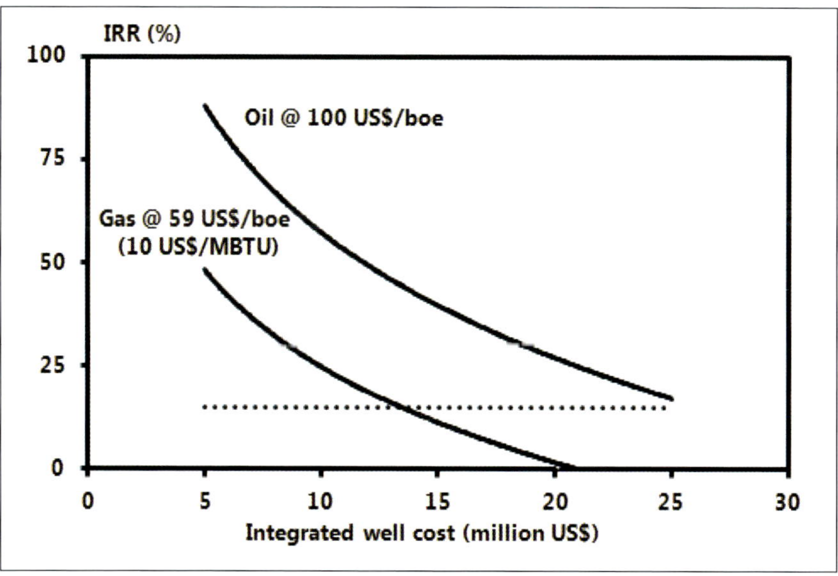

Figure 10 – Comparison of the economic profitability of oil and gas. Oil projects stay profitable even with high well costs, unlike gas projects which rapidly become uneconomic (Source Total).

Why have shale oil and gas caused such an outcry? Nuisances, risks and myths.

Any object, project, action, process or person represents a potential hazard. Accepting a hazard means accepting the notion of a risk which is associated with a potential incident, a severity level and a probability of occurrence. Taking a risk means having enough confidence to accept the probability of occurrence. The immediate consequence of losing confidence is the refusal of the probability of occurrence. The hazard then stops being a risk and becomes a threat. Unlike a risk, which is a rational relationship with a hazard, a threat is an irrational approach to a hazard. Three main reasons can tip the balance: facts, manipulation and vocabulary.

Over the last thirty years, few subjects have generated as many column inches, as much debate, aggression and accusations as those of shale oil and gas and hydraulic fracturing. Since this socio-media hurricane has not been whipped up from real proven health, safety and environment issues, the real reasons for the outcry must be sought in the other factors that trigger a menace: manipulation and vocabulary.

The film *Gasland* and its shocking image of the "*faucet on fire*" had a stunning effect on its viewers. Even though the scene was filmed in a region where shale gas has never been exploited, the lie doesn't matter. The allegory spread like wildfire, arousing a clamor of indignation from the general public and firmly planting in the collective imagination the idea that hydraulic fracturing is an environmental threat that pollutes all the water-bearing layers of the planet. Hydraulically fracturing the source rock also conjures up in the imagination of the layman, a full-on threat (rupture of a solid base rock, the Earth opens beneath our feet) but also an insidious one (a fluid brimming over with dangerous chemicals injected underground will seep out one day when we are least expecting it).

Even if the threat related to shale oil and gas has essentially been built up around suggested impressions and a poorly chosen vocabulary, the risks associated with development and production activities must not be denied. To give a reasoned explanation of where to draw the line between reality and fiction, issues are classified in three categories: nuisances, which are certain events of low severity, risks in the strict sense and the myths associated with a high severity level, but with an extremely low probability of occurrence.

The main nuisances that require mitigation are the impact at the surface, the greenhouse gas emissions generated by operational activities, water requirements and the possible conflicts in terms of usage and the production of waste. The main risks to be reduced are road accidents (related to truck traffic), any surface pollutant spills and possible well integrity problems.

The main myths that need to be dispelled are the contamination of water-bearing layers in the wake of an uncontrolled reflux of a hydraulic fracture, earthquakes and fugitive methane emissions caused by uncontrolled leaks at the well head or in the gas pipelines.

Hazard, risk and threat[1]

Any object, project, action, process or person is a potential **hazard**. This applies for example to a computer or a telephone (they use electricity and can cause electrocution), any means of transportation (car, train, boat or aircraft accidents), the construction of a building (which could fall down), a bank (it could go bankrupt) or to foodstuffs (which could cause food poisoning).

Accepting a hazard means accepting the notion of a **risk** which, associate with a potential incident (i.e. a plane crash), its severity (also called *criticality*) and the probability of it happening (we usually talk of probability of occurrence). Therefore a plane crash is associated with a high criticality (all the passengers are in danger of being killed) but a very low probability of occurrence (the number of planes that crash compared with the number of flights per day is very low). By boarding a flight, we take the risk which is very low, but nonetheless existent, of being on a

1. A. Mergier et G. Biasini (2014), "Opinion publique et danger. Comment parler de risques" http://www.palomar.fr/

plane that may crash. So flying by plane involves a certain amount of confidence in the type of plane and the airline so that we can accept the probability of occurrence, i.e. that an accident may happen.

If confidence is lost, the immediate consequence is the refusal of the probability of occurrence. The hazard then stops being a risk and becomes a threat. Refusing the probability of occurrence, even if it is very low, is equivalent to hiding behind the principle of precaution. The tendency is then to ignore any scientific reasoning and, paradoxically, it is what non-scientists say that becomes credible. The threat is experienced as an absence of proof of the absence of a hazard. Unlike a risk, which is a rational relationship to a hazard, a threat is an irrational approach. In the risk approach, the individual is the subject of the hazard, but when a risk becomes a threat, he becomes the object of the hazard.

Three main reasons can tip the balance: **facts**, **manipulation** and **vocabulary**.

More than anything else, it is a fact or an incident (the probabilized event actually occurs) of a given gravity which sways individual or public opinion from risk to threat. So the death of a close friend or relative in a plane crash, the bankruptcy of a bank, a terrorist attack in the subway or pollution that directly affects economic activity are triggering factors. Similarly, over the last twenty years, financial crises (subprimes, sovereign debt), ecological disasters (Macondo, Fukushima) or health scares (mad cow disease, bird flu) have significantly contributed to tipping the balance from risk to threat.

Catalyzed by the easy, instantaneous access to information (one news item replaces another!) in the media and on the Internet, hasty judgments lacking in substance and evidence, emotional accounts, shocking images, alarmist declarations, misleading claims and incomplete information have swamped the actual truth. The manipulation of information generates an irrational perception of the real situation, far removed from the actual facts. Little by little, the myth becomes a reality and takes root in the collective imagination. More often than not it is myths, rather than facts, that cause a risk to become a threat.

Beyond actual facts or myth-based manipulations, the vocabulary used by an expert can cause anxiety for the layman and trigger the switch from risk to threat. From the words he hears, the layman will build up his own imaginary knowledge. Restoring confidence also relies therefore on an alignment between the technician's language and that of the layman. So for example, the layman will associate GMO (Genetically Modified Organisms) with a genetic threat; a child who ingests GMO would become a monster.

The absence of facts

Over the last thirty years, few subjects have generated as many column inches, as much debate, aggression and accusations as those of shale oil and gas and hydraulic fracturing. This socio-media hurricane could have been based on proven incidents – health, safety environment – the consequences of which would have justified rejection by the stakeholders directly concerned. However, those who live in the regions where shale gases are produced are generally very much in favor of the project. The testimony given by Amy RUSTELEDGE, Director of the Chamber of Commerce in Carrollton, a small town in Ohio[2] is unequivocal. She explains[3]: "*Hydraulic fracturing is not a new technique even if everybody claims it is. Hydraulic fracturing has been going on in Ohio since 1950. For 60 years now, most of the wells have been fractured and there has never been any problem of polluted ground or water. I am 53 now and my family has been living in this small town for five generations. I would never accept to see them suffer.*" And yet in Europe, where there is no industrial production of shale oil and gas, hydraulic fracturing is criticized and considered as a threat that causes severe environmental damage, contaminating drinking water layers[4], releasing even more greenhouse gases than coal[5] and causing serious earthquakes[6]. When there is no backing proof, the real reasons for the outcry must be sought in the other factors that trigger a threat: manipulation and vocabulary.

2. Region of intensive production of Utica shales.
3. Triangle 7 RTBF "Made in Belgium» http://www.rtbf.be/info/economie/detail_quel-avenir-pour-notre-industrie-en-belgique?id=8242281
4. See question 16.
5. See question 18.
6. See question 17.

The manipulation of the "Gasland" film

On January 24, 2010, Gasland[7] the documentary by Josh Fox was screened at the Sundance independent film festival[8] in Utah and then a few months later, it was broadcast on the NBO TV channel.

The film had a stunning effect on its viewers and the images spread like wildfire among the general public. J. Fox had been approached by an oil company that wanted to drill wells on his family property in Pennsylvania in the Marcellus shales. His suspicions were aroused and he began to investigate the shale issue, piecing together a case to impugn the motives of all shale oil and gas producers and highlighting in particular the controversial issue of water pollution in the region. But it was the image of the faucet on fire (**Figure 1**) that went round the world at lightning speed, went viral on the social networks and put J.Fox in the spotlight. However, doubts soon began to emerge regarding the accuracy of the

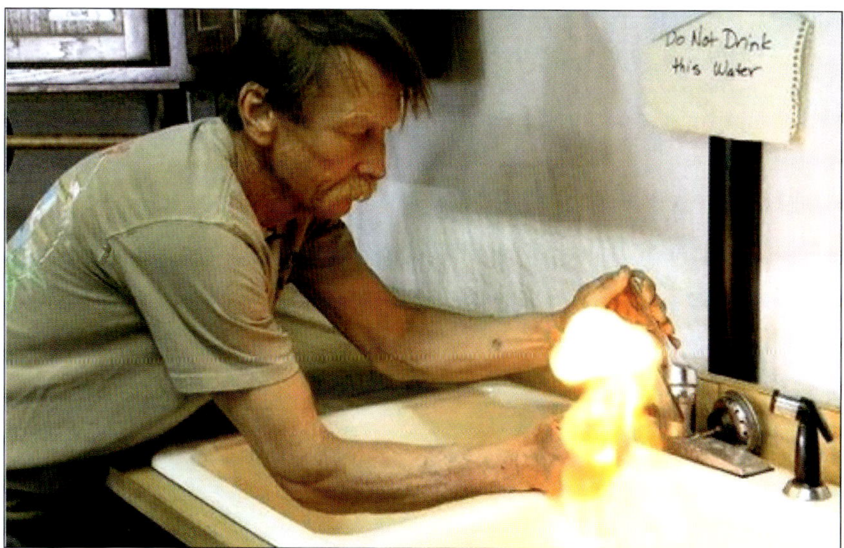

Figure 1 – Gasland: the image of the faucet on fire ignited the collective imagination and turned the risk of shale gas and oil into a threat.

7. http://gaslandthemovie.com/
8. http://www.sundance.org/festival/

image. Phelim Mc Aleer[9,10,11] a journalist from the Financial Times, followed the trail back to where the image was shot and discovered that it had been filmed on the border between Ohio and Indiana, in a region that has never seen any shale gas production. But the damage was done and regardless of the lie, the allegory fired the indignation of the general public and firmly anchored in the collective imagination the idea that hydraulic fracturing is an environmental threat that will contaminate all the drinking water resources on the planet.

The ontological danger of hydraulic fracturing

When a layman hears that to produce shale oil and gas we have to hydraulically fracture the source rock he imagines from these few words a full-on insidious threat. Full-on because fracturing is associated with breaking up a solid base rock, an earthquake during which the Earth opens beneath our feet. In addition, the use of hydraulic techniques is thought to deprive the human species of the very water which is the source of its existence. Yet hydraulic fracturing is also a sneaky insidious threat, as it uses a fluid brimming over with dangerous chemicals which is injected underground. Just like the gas in the water running from the faucet on fire, the chemicals will seep out one day when we are least expecting it. Finally, the technique is also immoral and brutal as it damages the source of life. Fracturing the source rock for the layman is perceived as an ontological threat[12] in the same way as nuclear war was perceived during the cold war. Faced with an ontological threat, there is no compensation that can offset the anguish except that of sheltering behind the principle of precaution.

9. http://www.youtube.com/watch?v=2cKhY2Edt2Q&list=PLE2AB13894C2961B7
10. http://www.youtube.com/user/noteviljustwrong
11. http://vimeo.com/user15547788/review/85615804/e793a10948
12. Ontology is a branch of philosophy which concerns the study of the being as a being, that is, the study of the general properties of everything that exists. An ontological risk therefore endangers the being in his existence. The environmental issue is an ontological hazard by its very nature, as it affects the survival of the human species

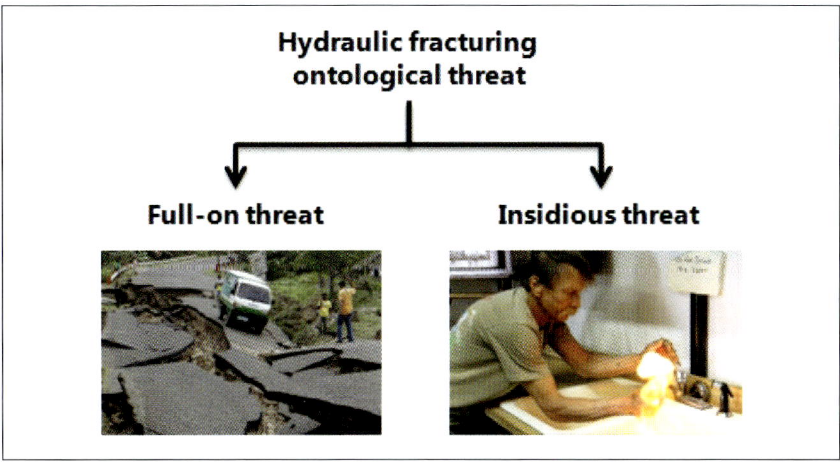

Figure 2 – The impact of words on perception. Hydraulic fracturing is both a full-on ontological threat and an insidious ontological threat

Identifying and classifying issues

Even though the threat related to shale oil and gas has essentially been built up around suggested impressions and not hard facts, the risks associated with development and production activities must not be denied. Restoring a capital of trust requires the issues to be explained in total transparency and to teach people to make the difference between hard facts and sensational rumors. To achieve this, the issues likely to cause environmental or social incidents are classified into three categories (**Figure 3**): nuisances, which are certain events (100% probability of occurrence) but with a low criticality, risks in the strict sense and the myths associated with a high criticality but with an extremely low probability of occurrence.

Nuisances, risks and myths are all treated differently. The criticality of nuisances is mitigated; risks are reduced by acting both on their probability of occurrence and their criticality. As for myths, which in public opinion are often the main threat, they must be dispelled using demonstrative measures, an open, instructive communication strategy and a language that can be understood by all.

Once all the issues have been identified and classified, a baseline must be established within the project perimeter to distinguish the potential

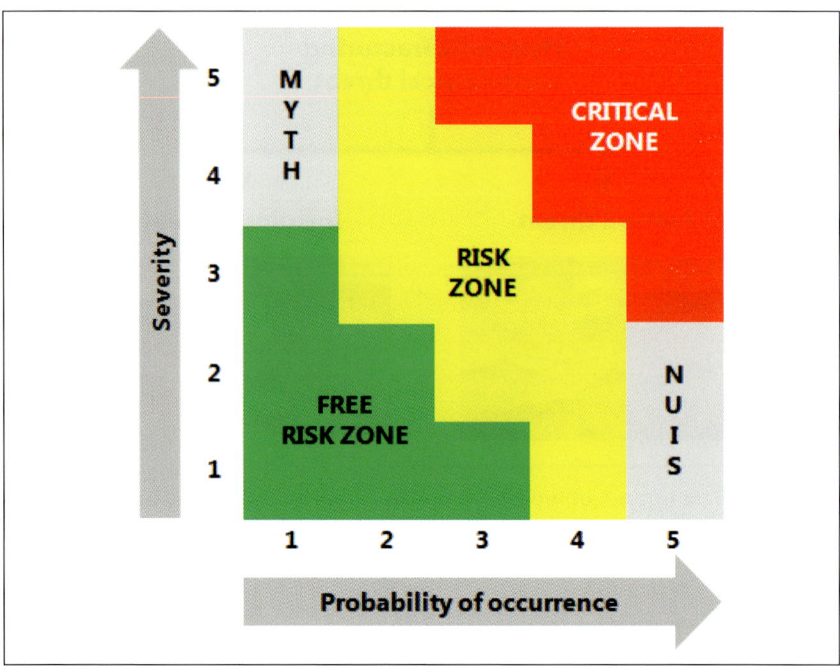

Figure 3 – In the diagram comparing criticality/probability of occurrence, nuisances (low criticality, high probability of occurrence) are placed in the bottom right-hand corner and myths (high criticality, low probability of occurrence) in the top left-hand corner. The central part of the diagram covers risks in the strict sense. The top right-hand corner is critical whereas the bottom left-hand corner is risk free.

risks from the issues that may arise from existing activities (domestic, industrial, agricultural). In most producing countries, this environmental baseline study is a legal requirement. It is usually carried out by a third party and covers many different aspects such as the cartography of the flora and fauna, seismic activity, natural radioactivity, air and water quality, ground conditions and any contamination inherited from former industrial projects. The social baseline study, carried out at the same time, is designed to map out the socio-economic context: demography, health, employment, housing, transportation. It also covers the map of stakeholders, in particular the local communities that are directly affected by the nuisances. Anticipating their perceptions is essential, since a crucial issue for one group of people may be secondary for another. The baseline study is followed by an impact study, also a legal requirement and designed to quantify the incidence of project activities relative to the baseline study. All the issues covered in the impact study will be continuously monitored throughout the production period using an appropriate measuring system.

The nuisance/risks/myths matrix specific to shale oil and gas

The main health, safety and environmental issues presented in **Figure 4** have been divided into nuisances, risks and myths.

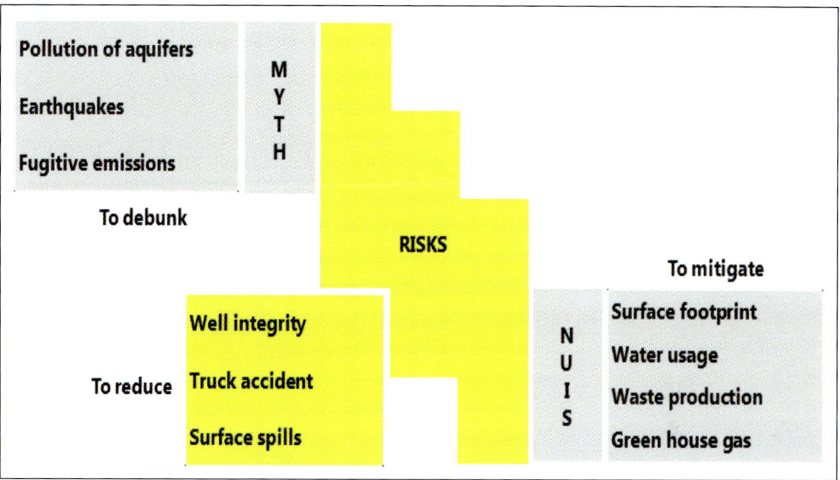

Figure 4 – Nuisances, risks and myths associated with the development of shale oil and gas.

The main nuisances that must be mitigated are:

- The surface impact[13] of operational activities (surface footprint, truck traffic, road damage, eyesores, noise and olfactory pollution related to operations, light, dust).
- Greenhouse gas emissions generated by exhaust fumes or discharge when a well is cleaned up[14].

13. This issue will be discussed in question 13.

14. This issue will be discussed in question 18.

- Water requirements and possible conflicts regarding usage[15].
- Waste, in particular the recovered fluid[16] which has to be treated.

The main risks to be reduced are:

- Road accidents (related to truck traffic).
- Any surface pollution caused by an accidental spill (on site or as a result of an accident involving a truck carrying non-ecological products - e.g. chemicals).
- Well integrity issues[17].

The main myths to be dispelled are:

- The pollution of water-bearing layers as a result of the uncontrolled migration of a hydraulic fracture[18].
- Earthquakes[19] caused by hydraulic fracturing.
- Fugitive methane emissions caused by uncontrolled leaks from the well or from the pipelines carrying the gas to the processing plant.[20]

15. This issue will be discussed in question 14.
16. This issue will be discussed in question 15.
17. This issue will be discussed in question 16.
18. This issue will be discussed in question 16.
19. This issue will be discussed in question 17.
20. This issue will be discussed in question 18.

Is the surface impact really so extensive? Strategies for mitigating surface impacts

The location of the well and surface installations has a certain physical footprint and the associated activities (drilling and fracturing) create nuisances.

Limiting surface impact requires first and foremost a reduction of the footprint. The oil and gas industry, the first to be concerned by such issues, developed the PAD concept, which entails drilling several horizontal wells (between 5 and 15) from a surface area equivalent to two to three football pitches. So, for example, a field that requires 1,000 wells can be developed using less than a hundred PADS. Although the PAD is particularly crowded during the drilling and fracturing phases, space is soon freed up in the production phase. The PAD has a high energy efficiency. To produce the equivalent quantity of electricity, it requires seven to ten times less surface area compared to solar panels and wind turbines. In addition to the PAD concept, reducing the physical footprint of installations also relies on optimizing the number of wells by identifying the sweet spots that could limit the number of non-productive wells. Reducing the footprint always goes hand-in-hand with economic profitability.

If we analyze the issue on a large scale it becomes apparent that the problem is essentially regional and not global. The aggressive development of shale oil and gas in Europe would have a footprint equivalent to the area of... Lake Geneva.

Limiting the surface impact also means reducing noise pollution, unpleasant smells and eyesores. The activities, though temporary, are associated with road traffic which is an intense source of dust, light pollution, greenhouse gases and road damage. Operational activities also affect biodiversity – the construction of PADs, truck traffic, water extraction and pipeline laying disrupt the habitats of numerous animals.

The main way of hiding an eyesore is to use the natural topography to best advantage. Building acoustic walls will reduce noise pollution. Electric drilling rigs directly connected to the networks should be used in preference to those that run on diesel. A detailed traffic plan should be drawn up for road traffic, with very strict rules in terms of speed and night usage, particularly in urban areas. Finally, fluids should be transported by pipeline rather than by road whenever possible.

Yet, whatever the mitigation measures may be, it is impossible to erase the surface impacts completely. In addition to these measures, oil and gas companies are therefore obliged to set up procedures to compensate for permanent or temporary negative effects that have not been corrected and, for obvious reasons, compensation in kind is preferable to financial compensation.

Any industrial activity has a surface impact revealed on the one hand by a physical footprint and on the other by a number of nuisances (noise, odors, light pollution, dust, road damage), and the development of oil and gas fields is no exception. The location of the well and surface installations (in particular the pipelines that carry hydrocarbons to the treatment center) has a certain physical footprint and the related activities (drilling and fracturing) generate nuisances.

Reducing the physical footprint

The first thing to do to limit surface impact is to reduce the physical footprint which, for an unconventional development, is roughly equivalent to the number of wells multiplied by the surface area of each well. It is by acting on these two factors that the ground coverage can be significantly reduced.

During the first half of the XXth century, conventional onshore developments (Azerbaijan, California, Texas, the Middle East, etc.) were made by drilling a large number of vertical wells close together. This method required vast areas of land and generated intensive nuisances such as eyesores, noise pollution and unpleasant smells, not to mention proximity risks for the local communities. The oil and gas industry, keeping these environmental issues to the fore, began to implement development techniques that were much more respectful of the environment and of the importance of not disrupting local communities.

For this reason the well-cluster concept, proposed offshore in the 1970s[1] was extended to onshore developments.

A PAD[2] involves drilling several horizontal wells (between 5 and 15) from a surface area equivalent to that of two or three football pitches, then connecting them to the same treatment plant. Compared with developments using vertical wells, this concept markedly reduces the physical footprint (**Figure 1**). So, for example, a field that requires 1,000 wells can be developed using less than a hundred PADs.

The life of a PAD is summarized in **Figure 2**. It begins with a land-leveling phase followed by the installation of a geotextile membrane to protect the ground and the subsurface from any accidental contamination, and then the construction of a noise abatement wall. Although the PAD is particularly crowded and creates eyesores and noise pollution during the drilling and fracturing phases, this is limited for the most part to the well, separators and storage tanks during the production phase.

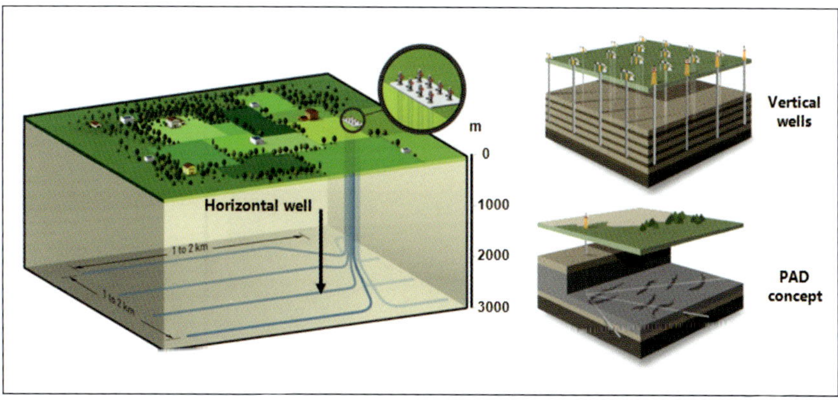

Figure 1 – The PAD concept markedly reduces the surface impact compared with older developments using numerous vertical wells.

1. See also question 6.
2. See also question 11.

Figure 2 – The life of a PAD starts with a land-leveling phase, followed by the installation of a liner and the construction of a noise abatement wall. The main nuisances are generated during the drilling and fracturing phases. During the production phase the PAD is reduced to a well area and a storage area and remains relatively uncluttered.

To highlight the energy efficiency of a PAD, in **Figure 3** we compared the surface area required to produce the same quantity of electricity[3] over twenty years for PADs of 10 conventional oil wells (3 Mboe per well[4]), shale gas wells (0.5 Mboe per well), solar panels[5, 6] and wind turbines[7]. The results show that the solar panels and wind turbines require 30 times the surface area that the oil PADs require and seven times more than the shale gas PADs. This is not surprising, in that for oil and gas the energy concentrated in the subsurface converges at one point at the surface, whereas for solar panels and wind turbines, the energy dispersed in the atmosphere has to be collected at the surface.

In addition to the PAD concept, reducing the physical footprint of installations also relies on optimizing the number of wells by identifying the sweet spots[8] to limit the number of non-productive wells inherent

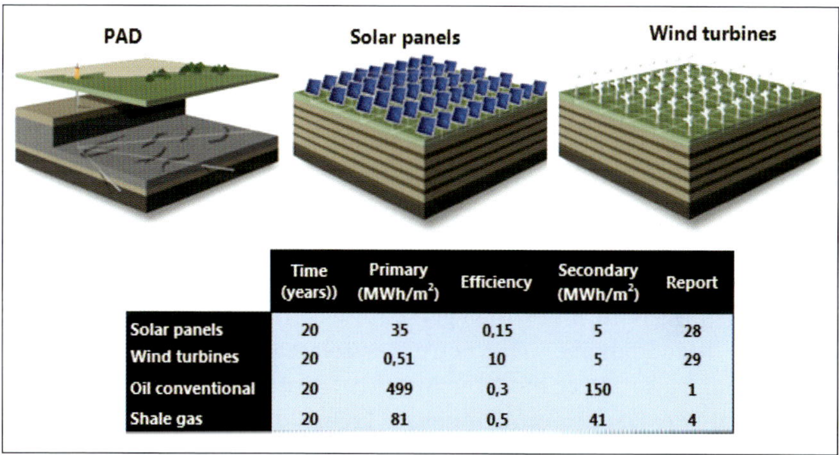

	Time (years))	Primary (MWh/m²)	Efficiency	Secondary (MWh/m²)	Report
Solar panels	20	35	0,15	5	28
Wind turbines	20	0,51	10	5	29
Oil conventional	20	499	0,3	150	1
Shale gas	20	81	0,5	41	4

Figure 3 – For the same quantity of electricity produced, a PAD has a much smaller physical footprint than solar panels and wind turbines.

3. The efficiency of heat cycles is 30% for oil and 50% for gas.
4. The efficiency of the steam cycle (30%) used by oil is lower than that of gas turbines (considered here as 50%).
5. 600 W/m², 8 hours of sunlight per day and a 15% yield for photovoltaic cells.
6. http://zebu.uoregon.edu/disted/ph162/l4.html
7. 627, 3MW wind turbines installed on a surface area of 100,000 acres (1 acre = 4047 m²), 15 hours of wind per day, 10% of the total surface used.
8. See question 11 Figure 5.

to the trial and error method used in the United States. The time taken to drill a well must also be reduced by utilizing the lightest possible well architectures, improving the penetration rate by the judicious selection of drill bits and drill strings, optimizing the time taken to install casing and carry out cementing operations - all without compromising on safety. It is crucial to note that any initiative that reduces the physical footprint will systematically contribute to economic profitability.

If we analyze the issue on a large scale it becomes apparent that the problem is essentially regional, as this type of development can come up against fierce opposition in densely-populated urban areas. However, on a global scale, the development[9] of shale oil and gas in the EU-28, which would require approximately 50,000 wells to be drilled, would occupy a surface area of no more than 500 km^2, i.e. a physical footprint equivalent to the area of Lake Geneva.

Attenuating peripheral nuisances

Limiting surface impact also means reducing noise pollution, unpleasant smells and eyesores[10], in particular those related to the different drilling and hydraulic fracturing operations. All these activities, although temporary[11], are associated with road traffic which is a source of dust, light pollution and greenhouse gases[12] (exhaust fumes from trucks and also from the gas turbines that generate electricity on site) and cause extensive road damage. The activities related to shale oil and gas development also affect biodiversity – the construction of PADs, truck traffic, water extraction and pipeline laying disrupts the habitats of numerous animals. The fauna is particularly affected by the noise, dust, light pollution and traffic, which must be managed appropriately to make sure that the natural environment is not damaged.

Perhaps the main way of hiding an eyesore is to use the natural topography (hills, forests) to best advantage but very simple modifications,

9. See question 20.

10. A drilling rig is approximately 35m high (compared with 50 to 100m for a wind turbine).

11. The PAD development phase (land leveling, drilling, fracturing, surface installations) lasts between 12 and 18 months, according to the number of wells, whereas the production phase lasts between 7 and 15 years.

12. See question 18.

such as directing the lighting on the drilling platform downwards, help reduce nuisances significantly. The construction of temporary noise abatement walls (around the PAD and the drilling rig - **Figure 2**) really dampens noise levels, which must be reduced to a maximum of 55dB during the day and 45dB at night. The new generation of electric drilling rigs that are directly connected to the network and therefore do not require generator sets, should be used in preference to diesel-powered rigs.

Regarding road traffic, a baseline study must first be drawn up specifying traffic intensity, nuisances (noise, light pollution, exhaust fumes), risks (accident occurrence rates) and the condition of the roads. Next, a detailed traffic plan should be drawn up in coordination with the authorities and local communities, with very strict rules in terms of speed and night usage, particularly in urban areas. Drivers must all receive appropriate training and heavy penalties shall be applied for infringement of the basic rules. Lastly, whenever possible, fluids (particularly water) should be transported by pipeline[13] rather than by road.

Compensating for any damage where mitigation measures were not sufficient

Whatever mitigation measures might be taken (e.g. repairing damaged roads), it is impossible to completely erase the surface impact. In addition to these measures, oil and gas companies are therefore obliged to set up procedures to compensate for negative effects that are permanent (i.e. areas of land occupied by PADs) or temporary (i.e. short-term loss of access to cultivated fields) and have not been corrected. Mitigation and compensation must restore and even improve living conditions for the communities affected. For obvious reasons, compensation in kind is preferable to financial compensation.

13. See question 14.

What is the water issue in shale oil and gas development?

Hydraulic fracturing operations seemingly require very large quantities of water. On average, a well with ten hydraulic fracturing stages requires 20,000 m^3, the equivalent of ten Olympic swimming pools. Yet when compared with water usage in farming, industry or for domestic purposes, such quantities remain modest. Producing electricity using shale gas instead of conventional gas only increases the quantity of water required by 1.7% for the entire cycle. To produce one MWh of electricity using shale gas, 16,000 times less water is required than if the ethanol produced from sugar cane were used.

Although the water requirements are negligible on a regional, national and global scale, they can cause conflicts in terms of usage with stakeholders on a local scale. To avoid any antagonism, the project manager must clearly state the sources, quantities and quality of water used, so as to reach an agreement with the authorities and with the local communities.

Hydraulic fracturing can easily adapt to using water that is unfit for domestic consumption or for agricultural irrigation. Near the coast, sea water can be used, or the slightly salty water from deltas and estuaries. The brackish water present in deep water-bearing layers can also be used, by contrast to the layers containing drinking water (also called groundwater layers) which are always located near the surface.

Since the water required for fracturing is rarely found close to the operations site, it needs to be carried over a certain distance to be stored on site. Instead of using road transport which can cause many kinds of disturbance for neighboring communities (noise, dust, light pollution, greenhouse gas emissions and above all, road damage), the alternative of pipeline transport is becoming more and more widespread today.

Whether transported by truck or in a pipeline, the water must then be stored on site. In most cases, it is stored in the open air in large artificial ponds dug on site, and lined with a geotextile membrane. Other types of equipment such as semi-rigid open top, or closed tanks, do away with the need for land-leveling works.

For certain arid or desert regions (Arabian Peninsula, North Africa or the Tarim basin in northwest China), supplying the required water can be a major obstacle. Alternative solutions using other fracturing fluids such as propane, may be more attractive in such cases.

Quantities in context

Hydraulic fracturing operations require large quantities of water (remember that the fracturing fluid contains 90% H_2O). In what follows, we will see that these volumes, which seem very large, must be put into perspective in comparison with other types of usage.

On average **one stage** of hydraulic fracturing requires approximately 2,000m^3 of water. A horizontal well with ten fracturing stages would therefore require around 20,000 m^3 of water, i.e. the equivalent of ten Olympic swimming pools[1] (**Figure 1**). Drilling is a lot less thirsty as it requires only 500 m^3 of water on average per well. When the well is brought on stream, between 25% and 50% of the volume injected will come back to the surface[2].

A mere drop of water on a regional or national scale

Water usage in a region or a country can be roughly divided into three main categories: 70% for farming, 20% for industry and the rest (10%) for domestic use (cooking, ablutions, cleaning, etc.). **Figure 2** shows the quantities of water required to produce a ton of different agricultural crops (maize, wheat, cotton, etc.) or a ton of certain metals (steel and aluminum). The 20,000m^3 of water needed to fracture a shale oil or gas

1. By way of comparison, an Olympic swimming pool contains 2,000 m^3 of water (50mx20mx2m).
2. Also called "*flowback*".

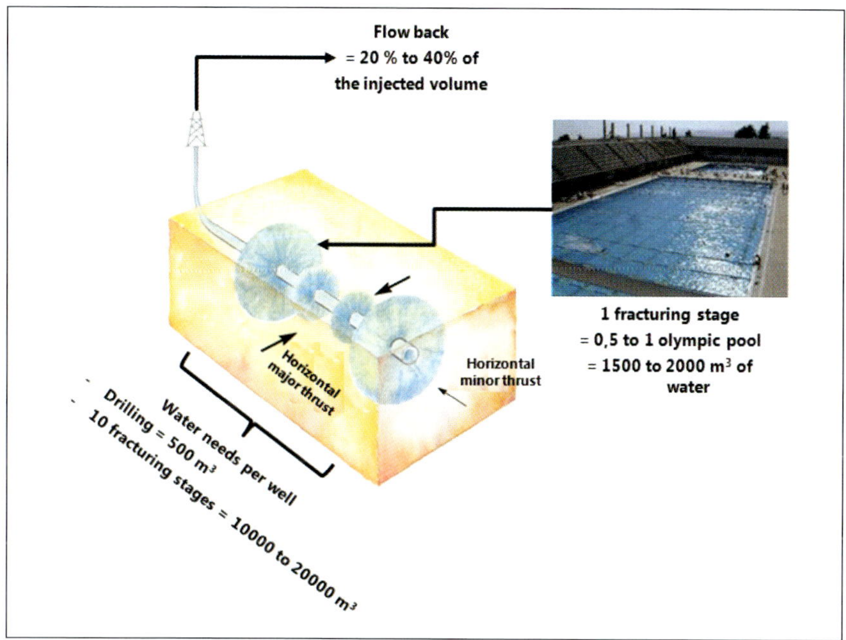

Figure 1 – Average quantities of water used to drill and multi-fracture a horizontal well.

well[3] would produce 33 tons of wheat, 22 tons of soya and just 4 tons of cotton[4] or rice. As for the production of metals, the same quantity is required to produce 250 tons of steel and 17 tons of aluminum.

If we take the same ratios, the 80 million tons of steel produced annually in the US would theoretically supply the drilling and fracturing of 320,000 wells, and the annual wheat production (about 60 million tons) would supply 2.4 million wells! The twenty thousand wells drilled every year in the US therefore represent a fairly modest quantity of water when compared with agricultural, industrial and domestic requirements.

3. Which assumes 100% loss, whereas we must remember that 25% to 50% is recovered as flowback.

4. The drying out of the Aral Sea is a consequence of the re-routing of the two powerful rivers Amou-Daria and Syr-Daria to mass produce cotton in Uzbekistan. It is one of the largest environmental disasters of the XXth century. http://fr.wikipedia.org/wiki/Mer_d%27Aral

The comparison with biofuels[5] is just as revealing. Producing a megawatt hour of electricity using shale gas requires 16,000 times less water than if the ethanol produced from sugar canes were used, which requires 275m^3 of water per MWh. If we compare the three main types of power generation, the gas power station, which directly uses the combustion of gas in a turbine, uses just one m^3 of water per MWh whereas nuclear or thermal (coal or oil-fired) power stations, which must first produce steam, need almost double that quantity (2m^3 of water per MWh of electricity). However, producing electricity using shale gas instead of conventional gas would mean an increase of just 1.7% in the quantity of water required (**Figure 2**).

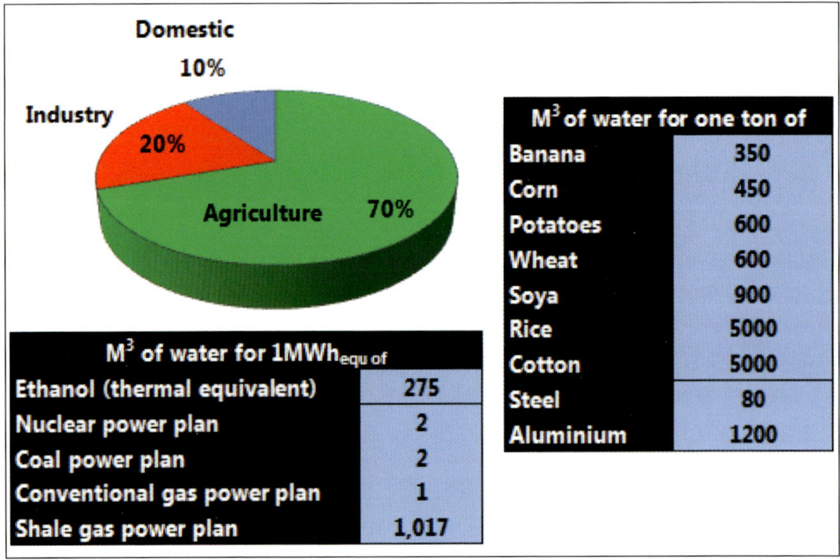

M^3 of water for 1MWh$_{equ\ of}$	
Ethanol (thermal equivalent)	275
Nuclear power plan	2
Coal power plan	2
Conventional gas power plan	1
Shale gas power plan	1,017

M^3 of water for one ton of	
Banana	350
Corn	450
Potatoes	600
Wheat	600
Soya	900
Rice	5000
Cotton	5000
Steel	80
Aluminium	1200

Figure 2 – Different water usages: farming[6], industry[7, 8] and energy production[9]. The comparison between shale gas and conventional gas shows that hydraulic fracturing requires only slightly more water to produce the same quantity of electricity.

5. Considering that sugar can requires irrigation.
6. CNRS (http://www.cnrs.fr/cw/dossiers/doseau/decouv/usages/consoAgri.html).
7. Water Use and Nuclear Power (Nuclear Energy Institute) (http://www.nei.org).
8. World Steel Association.
9. Ethanol equivalent energy is 26 103 MJ per ton (http://www.chimix.com/an7/prem/ethanol1.htm).

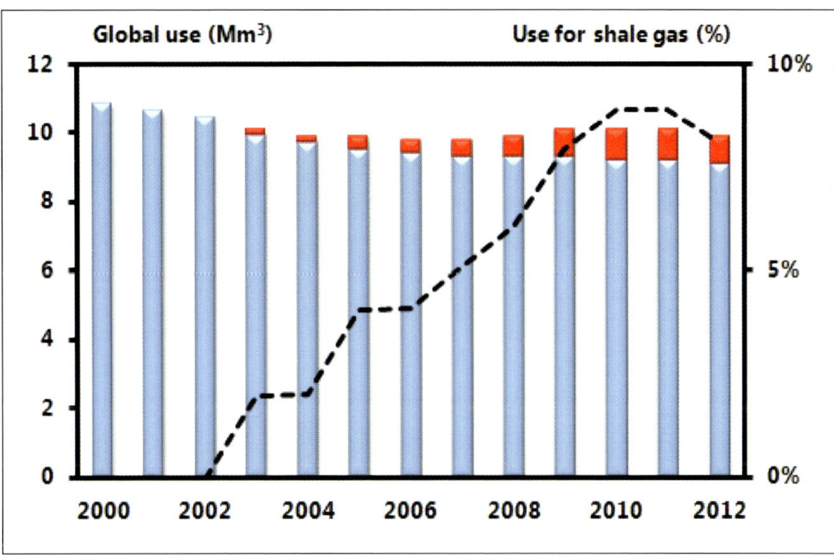

Figure 3 – Graph showing water use in the region designated for shale gas production in part of the Barnett field (Texas[10])

...but possible conflicts regarding local usage

Although the quantities of water required to fracture and produce source rocks may seem negligible on a regional, national and global scale, they may conflict with local agricultural, industrial or domestic requirements. So in 2011, in one of the regions within the Barnett field (**Figure 3**), the water extracted from a local ground water layer and used for shale gas production operations represented approximately 8% of the total volume of water extracted from the same layer.

No need for high quality water

Hydraulic fracturing does not require high quality water such as fresh water or drinking water. So to avoid any usage conflicts, water that is unfit for human consumption or agricultural irrigation can be used for shale gas production. Near the coast, sea water can be used, or the slightly salty

10. Source: Texas Water Development Board Report.

water from deltas and estuaries. The salt content[11] causes certain constraints in particular concerning the stability of certain chemicals such as thickeners, but these can be treated technically to overcome the problem. The brackish water found in deep water-bearing layers[12] can also be used, instead of the drinking water contained in the surface layers (also called *"ground water layers"*).

Selecting sources rationally and transparently

To avoid water usage conflicts and opposition from stakeholders, the water requirements and the sources used must be anticipated and communicated as early on as possible. The company responsible for the development project must make a clear commitment to the different stakeholders concerned (authorities and local communities) to reach an agreement on the **sources, quantities and quality** of water the project requires. This is done using a five-phase methodology:

- To start with, the profile of requirements (i.e. the volumes of water required according to time) must be defined for the different project activities. A typical example for the drilling of 90 PADS of 10 wells to maintain a production plateau of 250 Mcf/day[13] for 15 years is shown in **Figure 4.** The total quantity of water required is approximately 5 million m3, with a peak of 80,000m^3 at the start of the project. Regarding the proportions needed for the different activities, hydraulic fracturing will take up 95% of the water, drilling just 4% and the domestic requirements of personnel working on site represent less than 1%. In other words, and whenever possible, 99% of the requirements can be met using water that is unfit for human consumption and for irrigation. The estimate should be based on an average, reviewed periodically (either increased or decreased) in agreement with the stakeholders.

11. On average, sea water contains 30 grams of salt/liter. The chemicals and in particular the polymers contained in the fracturing fluid can tolerate a salt content of up to 50 of salt per liter of water.

12. The water held in the rock becomes more and more salty with depth, reaching a salt content of approximately 150grams of salt per liter of water, much higher than that of sea water. Remember that the maximum quantity of salt that can be dissolved in water is 230 grams per liter. Above this concentration, the salt is precipitated. http://www.captage-stockage-valorisation-co2.fr/aquiferes-salins-profonds

13. i.e. a sixteenth of the daily consumption in France is 4 Bcf per day.

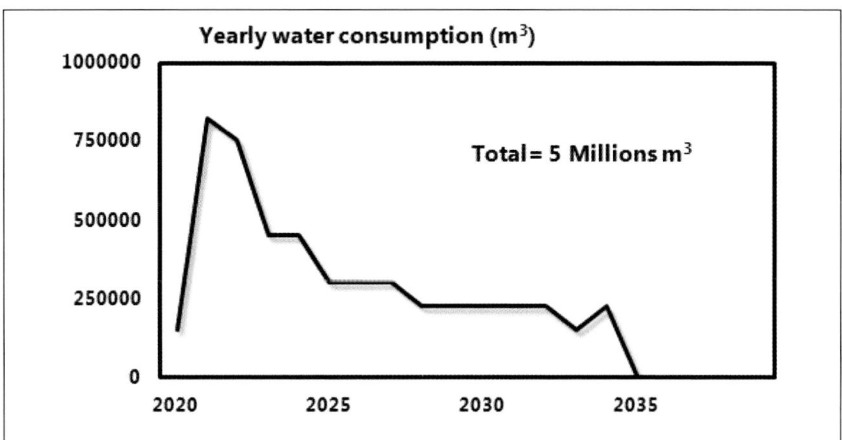

Figure 4 – Typical example of a water requirements profile. Hydraulic fracturing represents over 90%.

- Next, the quantity and quality of existing **sources** (ponds, lakes, streams, rivers, subsurface potable and non-potable water, estuaries, deltas and sea water in the case of activities located near the coast - **Figure 5**) must be identified and a **baseline** drawn up, which is both qualitative (list of consumers – domestic, agricultural, industrial) and quantitative (volumes drawn off by each user group). Governed by the water cycle (precipitation/run off/evaporation/condensation[14]), the level of these different sources heavily depends on the season (periods of flooding or drought) and the climate (temperate or arid regions). Since the different sources interact (e.g. subsurface water feeds into rivers and lakes and vice-versa), it is crucial to consider the entire regional water cycle by modeling circulation, exchange and storage. By incorporating the profile of requirements into this model, the hydrological **zone of influence** can be estimated for the project and its potential **impact** on the environment and users in terms of risks (negative impact relative to the baseline) and opportunities (positive impact relative to the baseline). In certain types of basins, the impacts can sometimes appear at a certain distance from the project.

14. See Appendix 8.

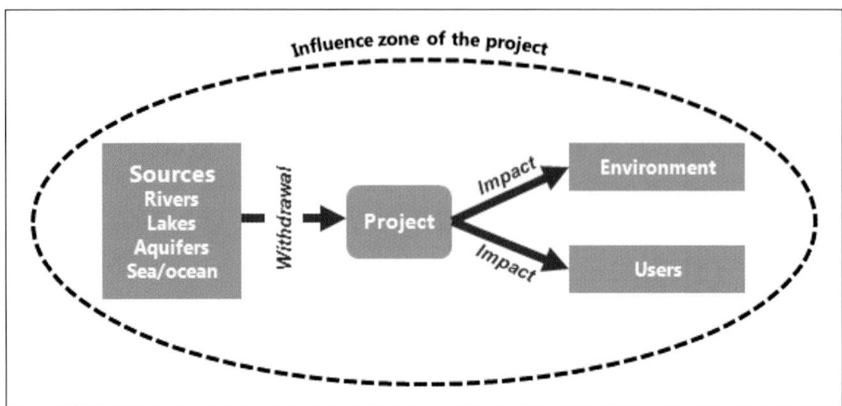

Figure 5 – Sources, hydrological influence area and impact.

- Different draw off methods which satisfy the requirements profile can then be compared to select in full transparency and above all, with the agreement of the main stakeholders affected, the water sources that will have the lowest environmental and social impacts (i.e. water use conflicts) while ensuring that the project remains profitable.

Transport and storage, a nuisance alleviated using innovative solutions

Since the water required for fracturing is rarely found close to the operations site, more often than not it needs to be transported over a certain distance (**Figure 6**).

Road transport means long queues of trucks, causing numerous nuisances (noise, dust, light pollution, greenhouse gas emissions[15] and road damage) for the people living close to the transport routes. The main disturbance is therefore not, contrary to what is often reported, the quantity of water used, but its transport from the source to the site. Given the number of trucks, this method of transport can also cause road accidents. The nuisances must be discussed openly with stakeholders (in particular the local communities) before the start of operations. A baseline study of the road network must be carried out and the operating company must undertake to repair or compensate for any damage caused.

15. The specific issue of greenhouse gases will be discussed in detail in question 18.

Figure 6 – Water transport in tankers is gradually being replaced by temporary overhead or permanently buried pipelines. They can either be rigid or wound onto reels for easy deployment.

The alternative to tanker transport is to carry the water in overhead pipelines installed temporarily to supply the sites from the sources. However, a long-term part of the network or any sections that are too visible, can be buried in trenches. A wide range of pipes made of different materials (steel, aluminum, PVC, reinforced rubber) are now available on the market. To facilitate deployment and reuse, piping wound onto large reels is also available.

Whether transported by truck or by pipe, the water must then be stored on the fracturing site. In most cases, it is stored in the open air in large artificial ponds dug out on site and lined with a geotextile membrane to prevent any infiltration into the ground underneath. Other types of equipment (semi-rigid open top or closed tanks - **Figure 7**) can be used to avoid the land-leveling works required in preparation for the artificial ponds. However, their storage capacity is much lower.

Figure 7 – Different methods are used to store water, either in the open air or in rigid or flexible tanks.

A blocking factor in certain regions

In certain arid or desert regions that hold vast shale oil and gas resources (Arabian Peninsula, North Africa or the Tarim basin in northwest China), supplying the water required for hydraulic fracturing can become a blocking factor. This also applies to subarctic regions (for example western Siberia) where temperatures remain negative for more than six months a year. In this kind of environment, alternative solutions using other fracturing fluids such as LPG, can be attractive.

Bringing a well on stream: what does flowback contain and how is it treated?

During the initial well production period called the flowback phase, the fluid recovered at the surface comprises essentially the injection water, some sand, a small quantity of residual chemical additives and a small quantity of rock broken off and either dissolved or in suspension. As the organic matter in certain source rocks contains radium, potassium, barium, uranium and thorium in small quantities, traces of these radioactive elements can also be found in the fluid.

As time goes on, the water content of the fluid produced decreases and the hydrocarbon content increases. However the flowback phase, during which the fluid recovered still contains injection water, can last over several months. In the end, only 20-40% of the volume injected will be recovered at the surface, the rest is permanently trapped in the source rock.

Former practices entailed cleaning up the well in the open air, into an artificial pond. This technique is open to criticism as the flowback fluid also contains dissolved gas which is therefore discharged into the atmosphere. Nowadays, to avoid these fugitive emissions of methane, wells are cleaned up directly into a separator.

Once degassed the liquid recovered, which in addition to water still contains many residues, is a waste product. The former method which was to directly discharge an untreated fluid into the environment is now strictly prohibited and two alternative techniques have taken its place.

The first involves slightly treating the fluid and then reinjecting it into a deep permeable formation through a dedicated well. Today, retreatment is preferred, because even if it is slightly more expensive, it is much more in line with environmental standards as the treated water can then be reused.

Treatment entails extracting the different types of waste (bacteria, oil or gas, solid particles in suspension, salt and other dissolved ions, possibly including traces of radioactive elements) using different purification techniques such as UV oxidation, gravity separation, filtration and flotation. At this stage, the treated water can be reused for the next hydraulic fracturing operation.

However, if the water is to be discarded on the ground, it must also be desalinated and cleansed of the other dissolved ions (in particular the trace of radioactive elements) using more sophisticated techniques such as evaporation or crystallization. Finally, if the water is to be discharged into the river, it must be distilled to extract any final traces of oil.

The residues from all these separation phases will be transported to a treatment plant to be stored and then destroyed, usually by incineration.

The fracturing fluid cycle

Supplied from different available sources[1], the water to be used for hydraulic fracturing is first stored on the fracturing site. It is then mixed[2] with sand (on average 9%) and chemicals (on average 1%) before being injected into the formation to be fractured. The flowback fluid recovered at the surface when the well is brought on stream can then be treated to be discharged or reused (**Figure 1**).

What does flowback fluid contain?

Once the operation has been completed, the fracture network, coated in sand, is essentially filled with fracturing fluid. During the initial well production period, called the "*flowback phase*", the fluid recovered at the surface (**Figure 2**) will contain mainly injection water, some sand, a small quantity of residual chemical additives and a small amount of rock leached and either dissolved or in suspension.

1. See question 14.
2. See question 8.

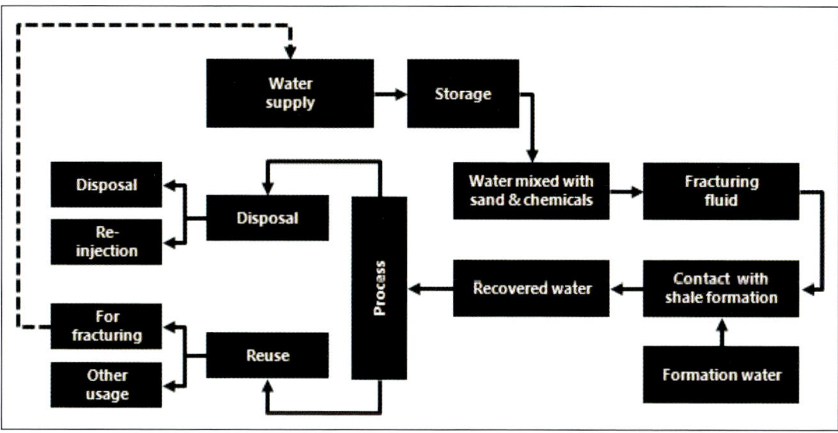

Figure 1 – The water cycle from supply to reutilization.

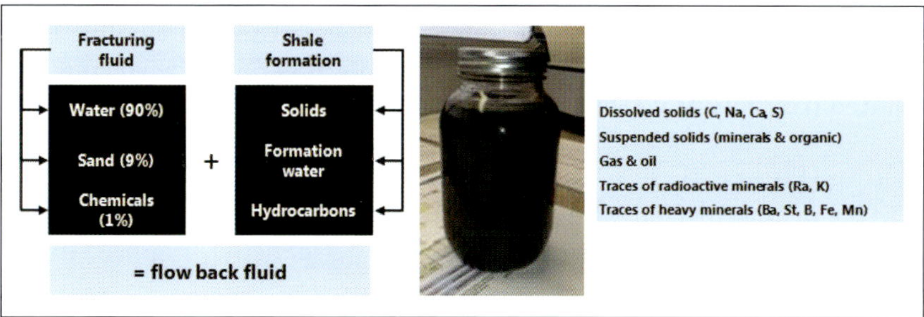

Figure 2 – Flowback contains essentially fracturing water but also small quantities of many different ingredients from the formation.

As the organic matter in certain source rocks contains radium, potassium, barium, uranium and thorium in small quantities, traces of these radioactive elements can also be found in the recovered fluid. According to a recent study performed by the CEREES[3], the radioactivity of the recovered fluid is however less than for conventional oil & gas. Over time, the water content of the fluid produced naturally decreases and the hydrocarbon content increases, but the period during which the fluid recovered will still contain injection water can extend over several

3. Almond S., Clancy, S.A., Davies, R.J., Worrall, F. (2014) "The flux of radionuclides in flowback fluid from shale gas exploitation" Centre for Research into Earth Energy Systems (CeREES), Department of Earth Sciences, 5 Durham University, Science Labs, Durham DH1 3LE, UK. See Appendix 9.

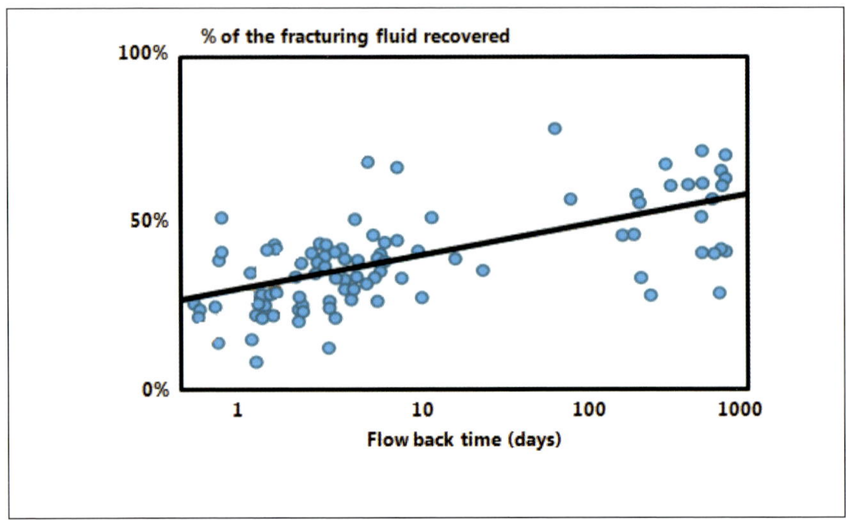

Figure 3 – The flow out of the fracturing fluid can last several months. The volume recovered at the surface represents only 20-40% of the volume injected (courtesy of GASFRAC & boe solutions).

months (**Figure 3**). Moreover, even after a very long period of time, only 20-40% of the injected volume will be recovered at the surface. The rest (i.e. 60-80% of the volume injected) will be permanently trapped[4] in the source rock.

How is flowback fluid stored?

Former practices entailed, for the first few days, flowing back the well in an open artificial pond lined with an impermeable membrane (a liner like those used in swimming pools or walk-in showers - **Figure 4**).

This simple low-cost technique, which prevents the fluid infiltrating into the ground, is however open to criticism when it comes to greenhouse gas emissions. Even though the initial flowback fluid is essentially water, it also contains dissolved gas. Just as when a bottle of sparkling mineral water is opened, when the flowback fluid arrives at the surface it will degas, releasing methane, a gas whose *"greenhouse gas potential"*[5] is 25

4. See question 11.
5. This issue will be discussed in detail in question 18 and Appendix 11.

Figure 4 – Similarity between the liner of a home swimming pool (left) and the membrane of an artificial clearing pond (right).

times higher than that of CO_2. Nowadays, to avoid these *"fugitive emissions"* of methane, wells are flowed back directly into a separator, a large horizontal barrel located at the well outlet in which the flowback fluid undergoes gravity separation. Once it has degassed inside the separator, the heavier liquid that settles at the bottom of the barrel can be recovered at the base outlet. The much lighter gas is drawn off at the top outlet.

How is flowback treated?

Once degassed the recovered liquid still contains many chemical residues and dissolved solids or particles in suspension, and is therefore a waste product. The old method in the 1970s and 80s (i.e. well before the shale oil and gas revolution) which involved discharging untreated flowback directly into the environment (e.g. into rivers) is now strictly prohibited by producing countries and by the best practice of international companies.

The first alternative technique involves not treating (or minimally treating) the flowback and then reinjecting it into a deep permeable formation (i.e. a non-potable water-bearing layer) through a dedicated well. Frequently used in conventional projects[6] and also in the United States in shale oil and gas projects, PWRI[7] carries three major disadvantages. To start with, even if the host formation is deep and contains only water that is unfit for any other usage (domestic, industrial or agricultural), we

6. Conventional projects produce increasing quantities of water over time. On average today, $3m^3$ of water are produced for $1m^3$ of oil.

7. Produced Water Re-Injection.

Remove	Treatment type						
	Oxydation	Separation	Filtration	Flotation	Evaporation	Crystallization	Distillation
Bacteria	✓						
Oil		✓	✓	✓			✓
Solids		✓	✓	✓			
Salt					✓	✓	
Other ions					✓	✓	✓
Treatment for	Re-use				Soil disposal		River disposal

Figure 5 – The different types of treatment used to obtain either a reinjection water that does not contain bacteria, oil or solids or discharge water (ground or river) that has been fully desalted.

are talking here about significant quantities of untreated effluents being injected into the subsurface layers[8]. Secondly, experience shows that PWRI can cause minor earthquakes[9]. Finally, injected water is permanently lost whereas, if it had been treated appropriately, it could have been reused. This is why retreatment is preferred today as, although more expensive, it is much more in line with environmental standards.

Treating flowback fluid involves removing the different types of waste (bacteria, oil or gas, solid particles in suspension and other dissolved ions, possibly including traces of radioactive elements) using different purification techniques[10] (**Figure 5**). UV oxidation removes bacteria, and gravity separation, filtration and flotation remove the solids and oil in suspension. The treated water can then be used for the next hydraulic fracturing operation.

To desalinate the water and remove the other ions dissolved in it (in particular radioactive elements) much more sophisticated and more expensive techniques are required, such as evaporation or crystallization. Finally, if the water is to be discharged into the river, it must be distilled to extract any final traces of oil in particular.

The residues from all these separation phases will be transported to a treatment plant to be stored and then destroyed, usually by incineration.

8. See also question 17 on groundwater pollution.
9. See question 16 on seismic activity.
10. See appendix 10 for more information.

Can hydraulic fracturing cause earthquakes?

Earthquakes are caused by the sudden slippage of large, deep fractures. They are distributed along narrow converging tectonic plate margins. The intensity of an earthquake is measured on the Richter scale, which also indicates the consequences. Only earthquakes higher than 3 can be felt by humans and damage is observed for intensities greater than or equal to 4. Over a million earthquakes are recorded worldwide each year, about 2,000 of which have an intensity greater than 5.

Although most earthquakes have natural causes, certain human activities such as mining excavation or the variations in water levels in a hydraulic dam, can cause minor seismic events. This phenomenon is referred to as induced seismicity. Similarly, hydrocarbon production, water injection and hydraulic fracturing induce very minor seismic events. Those that occurred in Blackpool in England, with an intensity of between 1.5 and 2.3, released energy levels 5 to 30 times lower than those released by the vibrations of a train running in the Paris subway.

When a microseismic event occurs, the energy released emits noise which travels through the surrounding rock as sound waves. Measuring them requires highly sensitive detectors. This method, called microseismic monitoring, locates microseisms events in space and contributes to mapping the fractured area around the well. Micro-seismic detection associated with hydraulic fracturing is also an opportunity to measure (and therefore to control) the propagation of the fractured area.

Unless a field is developed in a highly active tectonic region, the seismic risks related to hydraulic fracturing are extremely low. Nonetheless, for stakeholders they represent an ontological threat that must be given serious consideration.

Any shale oil and gas development project must include a natural seismicity baseline study for the zone concerned. It must include the history, frequency and intensity of natural and induced seismicity. Next, an impact study should be carried out, including digital simulations to estimate whether or not the seismic activity generated by the project could be higher than existing levels and likely to interfere with human, social and economic activities.

In order to monitor seismic activity continuously, the area concerned by the project will be covered by a monitoring network to detect seismic events of an intensity less than 1 on the Richter scale.

Origin and classification of earthquakes

Earthquakes are caused by the sudden slippage of large, deep faults (several dozen or sometimes several hundred kilometers deep) that instantly release huge quantities of energy. Their distribution across the globe is not random, but rather follows narrow fault lines along tectonic plate margins (**Figure 1**).

The intensity of an earthquake is graduated on the Richter scale (Rs), which is logarithmic (**Figure 2**). In other words, when the unit is increased by one, the intensity of the earthquake is multiplied by ten. The Richter scale also links the intensity of an earthquake to the consequences of such an event. Below 3, earthquakes cannot be felt and only high-precision instruments can record them.

Seismic events higher than three can be felt by humans (this is the intensity of an earthquake induced by a train running on the Paris subway) but no damage is observed unless the intensity is greater than or equal to 4. Over a million earthquakes are recorded worldwide (especially at sea) each year, about 2,000 of which have an intensity greater than 5 and the number of such earthquakes appears to have been increasing over the last decade[1] (1,500 in 2000, 2,000 in 2010). The largest earthquake ever recorded hit 9.5 on the Richter scale in Chile on May 22, 1960. Those in Sumatra in 2004 and Japan in 2011 (Fukushima disaster) were offshore earthquakes that registered 9 on the Richter scale and induced gigantic tsunamis. It is this kind of event and not earthquakes themselves that claim so many victims.

1. http://www.wikistrike.com/article-le-nombre-de-seismes-en-augmentation-83624291.html

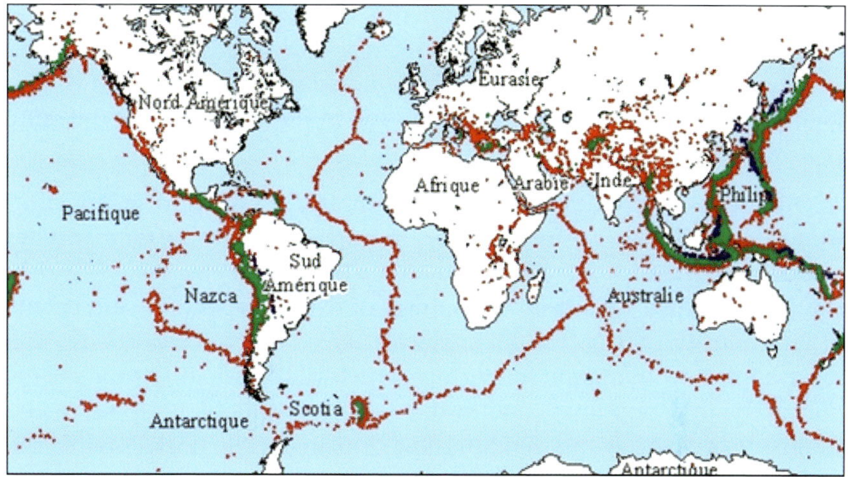

Figure 1 – Global seismic activity begins at a depth of 70-350 km on the tectonic plate margins, akin to lines of stitching between portions of the earth's crust. The map represents earthquakes of an intensity higher than 5 on the Richter scale. (Source: Institut de Physique du Globe, Paris)

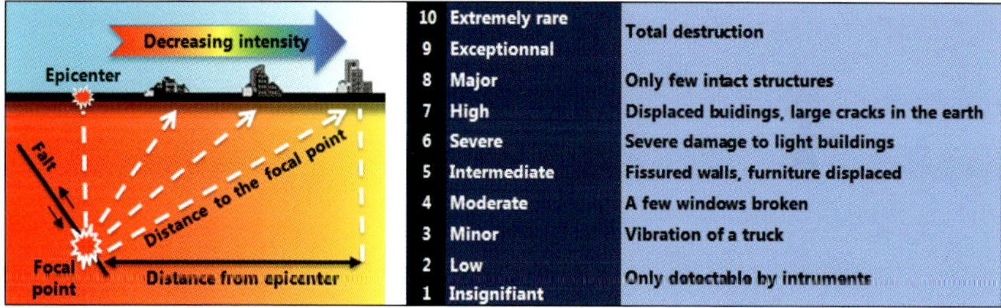

Figure 2 – Earthquakes are caused by the slippage of major faults. Logarithmic Richter scale.

Induced seismicity

Although most earthquakes have natural causes, certain human activities such as mining excavation or the variations in water levels in a hydraulic fill dam can cause micro-slippage in the encasing rock, equivalent to very small seismic events. They are mostly lower than 3 on the Richter scale and the phenomenon is referred to as *"induced seismicity"*. Extracting or injecting fluid into the ground can also induce seismic events. For example, injecting water[2] into hot dry rocks to produce geothermal energy caused moderate seismic events in Soultz[3] (eastern France, 2.9 Rs) and near Basel[4] (Switzerland, 3.4). Exceptional values exceeding 4 on the Richter scale were recorded in Australia and in California during the same type of operations.

Oil and gas activities are no exception and production or injection induce very minor seismic events. In particular, the seismic surveys associated with the extraction of hydrocarbons from the Lacq[5,6] field in southwest France have been the subject of extensive studies, and events of an intensity of 3 on the Richter scale have been recorded there. The same applies to hydraulic fracturing[7] which causes small seismic events during the injection phase and in the well flowback phase. However, the intensity of such events remains low and in fact it is essentially the reinjection of production water[8] and not hydraulic fracturing which was the cause of a certain number of seismic events higher than 2.5 Rs in the United States. Those recorded in Blackpool[9] in England (**Figure 3**) which "*shook the town*" according to the press, in fact were of intensities between 1.5 and 2.3 on the Richter scale, an energy level 5 to 30 times lower than that released by the vibrations of a train running on the Paris subway.

2. Mark D. Zoback (2012), Managing the Seismic Risk Posed by Waste water Disposal - www.earthmagazine.org
3. http://www.brgm.fr/brgm
4. http://www.liberation.fr/terre/010191741-la-geothermie-fait-frissonner-la-suisse
5. Bardainne, N. Dubos-Sallée, G. Sénéchal, P. Gaillot and H. Perroud (2008), "Analysis of the induced seismicity of the Lacq gas field (Southwestern France) and model of deformation" Geophys. J. Int.
6. JR Grasso, "Effet des injections sur la sismicite du champ de Lacq" (1975 - 1995)
7. http://gallery.mailchimp.com/27ed33877e4cbef4f0b684b7b/files/FracQuake_Backgroun-der_Ao_t_2012_4_.pdf
8. http://www.actualites-news-environnement.com/27717-fracturation-hydraulique-seismes.html
9. http://schiste.owni.fr/2011/06/01/seism-a-cause-de-la-fracturation-hydraulique-en-angle-terre-gaz-de-schiste/

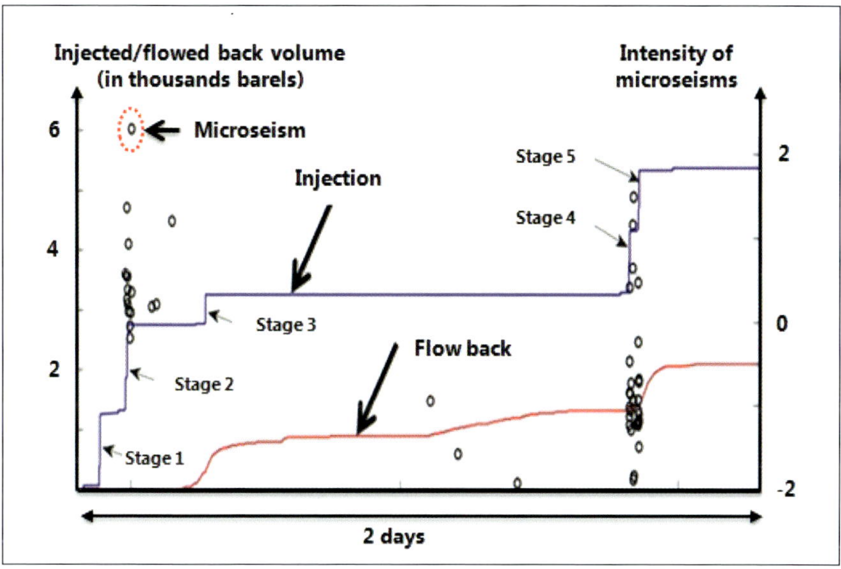

Figure 3 – Blackpool Shale Seismicity Hydraulic fracturing causes small seismic events reaching up to 2.3 on the Richter scale by relieving stresses on existing natural faults (Source: de Pater & Baisch)

How is induced seismicity measured?

When a micro-seismic event occurs, the energy released emits noise which travels through the surrounding rock as sound waves, called "seismic waves". Measuring such weak vibrations requires the installation of very sensitive detectors called geophones, either at the surface or in an observation well (**Figure 4**). This method also helps locate microseisms in space and therefore map the fractured area around the well.

Contrary to popular belief however, it is not the opening of the major fractures[10] (the highways[11]) but the slippage associated with the secondary fissures (the country roads) that generate most of the seismic activity during a hydraulic fracturing operation. Micro-seismic detection associated with hydraulic fracturing, is also an opportunity to measure, and therefore to control, the extension of the fractured area.

10. Opening mode 1 does not generate sound, it is *"aseismic"*.

11. See question 8, Figure 2.

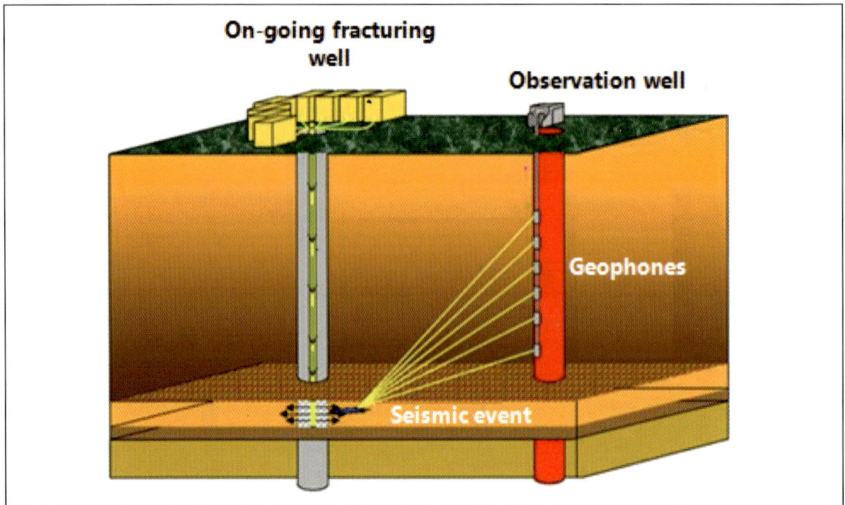

Figure 4 - Diagram of microseismic recording using geophones located in an observation well[12].

Two examples of microseismic surveys are presented in **Figure 5**. In the case of the Barnett shales, the well has nine fracturing stages (each color represents a different stage). The results of the microseismic recording show the development of a complex SRV (Stimulated Rock Volume) with very distinct fracturing stages (plan view) that are perfectly contained (vertical view) in the reservoir. In the Montney formation however, the microseismic data suggest the development of a system of planar, transverse fractures (highways with no secondary roads or small country roads!) not very conducive to a good production rate.

Seismicity as an integral part of the environmental baseline study

Unless a field is developed in a highly active tectonic region, the seismic risks related to hydraulic fracturing are extremely low. Nonetheless for stakeholders, they represent an ontological threat that must be given serious consideration.

12. http://www.harbourdom.de/frac_monitoring.htm

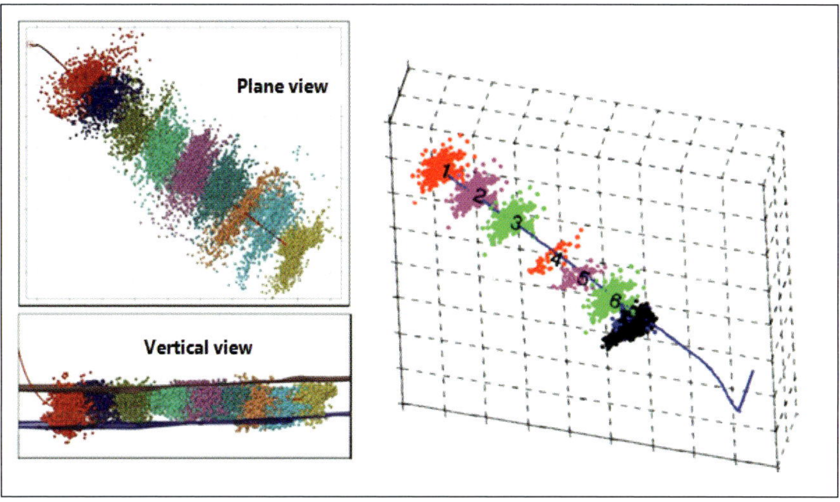

Figure 5 – Two examples of microseismic scatter graphs, each color represents a different fracturing stage. In the Barnett shales (left) the scattering shows a complex fractured volume with very distinct, well-contained fracturing stages. In Montney (right) the fractures are planar.

Any shale oil and gas development project must include a natural seismicity baseline study for the zone concerned. It must include the history, frequency and intensity of natural and induced seismicity. Next, an impact study should be carried out, including numerical simulations to estimate whether or not the seismic activity generated by the project could be higher than existing levels and likely to interfere with human, social and economic activities.

In order to monitor seismic activity continuously, the project zone will be covered by a monitoring network to detect seismic events of an intensity less than 1 on the Richter scale.

Is producing shale oil and gas a pollution threat for the groundwater layers?

The image of the faucet on fire shown in the *Gasland* film is founded on the idea that a hydraulic fracture, which initiates at a depth of several thousand meters, could make its way up to the surface, thereby opening a channel between the source rock and a potable water-bearing layer, allowing the fluids to flood in and finally end up in the water that runs out of the faucet.

The risk of groundwater layers being polluted by hydraulic fracturing must first be considered on an appropriate scale. Most shale oil and gas is located at depths of between 2,500 and 3,500m. Given the volumes pumped into the wells and the pumping power used, the vertical propagation of the fracture is limited to a few dozen or, in exceptional cases, a few hundred meters above the impregnated layer, whereas to reach the groundwater layer at the surface, it would have to travel a distance equivalent to the height of several Eiffel towers. All this is confirmed by microseismic monitoring techniques. Thousands of data have demonstrated that the vertical propagation of the fracture never exceeds 350m, so by taking a safety distance of approximately 750m between the zone to be fractured and the potable water-bearing layer, the probability of communication between the two can be considered nil. Even if the volume of fluid injected were to be increased infinitely, the fracture would have very little chance of propagating right to the surface. At depths of 750-1,000m, as the vertical thrust becomes very often lower than the horizontal thrust, a fracture that reaches these depths quite often turns and propagates horizontally. So the contamination of groundwater layers caused by hydraulic fracturing techniques is not feasible. After several million hydraulic fractures made worldwide over more than 70 years, no cases of groundwater pollution associated with hydraulic fracturing operations have ever been reported.

Although extremely rare, the percolation of a fluid along a well is however possible. But it is not specific to shale oil and gas. In theory, the cemented casing assembly creates a seal that allows the hydrocarbons to flow to the surface without ever coming into contact with the different layers they travel through. Over time, the cement sheaths can become damaged and let small quantities of fluid percolate through, but even then, the nesting of the different casing layers makes a leak into the surface groundwater layers highly unlikely. Only a single case of a leak to the surface has been recorded in worldwide shale oil and gas production, after a loss of well integrity.

The image of the faucet on fire shown in the *Gasland* film[1] is founded on the simplistic idea that a hydraulic fracture, which initiates at a depth of several thousand meters, could make its way up to the surface, thereby opening a channel between the source rock and a potable water-bearing layer. The fluids injected (fracturing fluid and the chemicals it contains) or formation fluids (oil and gas) from the source rock would therefore percolate through the fracture and flood into the potable water-bearing layer and finally end up in the water that runs out of the faucet. It has now been clearly demonstrated that the faucet on fire was a case of biogenic gas which had nothing whatsoever to do with the production of shale oil and gas.

The vertical propagation of a fracture…

The problem of vertical "containment" of hydraulic fractures is a problem that has been extensively researched in the oil and gas industry since the beginning of the 80s. The best analogy[2] consists in placing on a solid table a flexible sheet of metal held in position by two heavy stones at either end (**Figure 1**). Imagine that a hole is pierced in the center of the metal sheet and a metal tube welded into it, through which fluid is injected. The fluid begins to percolate between the metal sheet and the table in a roughly circular pattern, but when it reaches the area held down by the stones, it is much easier for the fluid to run along the x-axis than to lift the stones.

1. See question 12.
2. Ph. A. Charlez (1997), *Rock Mechanics. Vol II. Petroleum Applications*, Editions Technip.

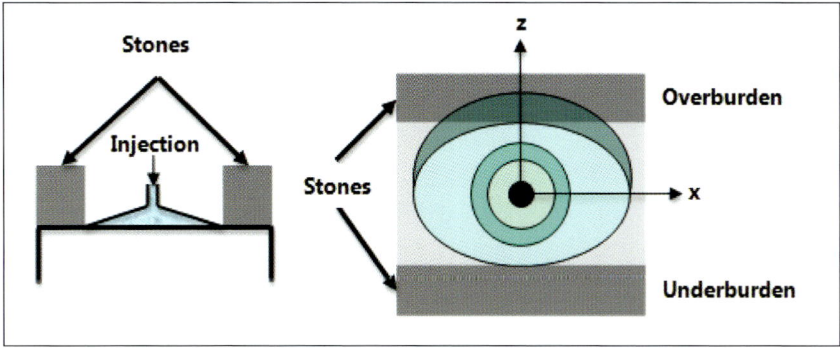

Figure 1 – Analogy to explain the action of horizontal thrust in caprock formations. The fracture propagates mainly along the x-axis and not the z-axis

In practice, the stones represent the horizontal thrust in the caprock formations (and in the underburden). Just as in the table analogy, they force the fracture to propagate more horizontally (along the x-axis) than vertically (along the z-axis). However, when the fracture has traveled for a long enough horizontal distance, it can be easier for the fluid to lift the stones than to continue to propagate along the x-axis.

... is more than anything, a question of scale...

The risk of groundwater layers being polluted by hydraulic fracturing must first be considered on an appropriate scale. Most shale oil and gas is located at depths of between 2,500 and 3,500m. Given the volumes pumped into the wells[3] and the pumping power regularly used, the vertical penetration of the fracture is limited to a few dozen or, in exceptional cases, a few hundred meters above the impregnated layer, whereas to reach the groundwater layer at the surface, it would have to travel a distance equivalent to the height of several Eiffel towers (**Figure 2**).

3. 2,000 m^3 per stage – see question 14.

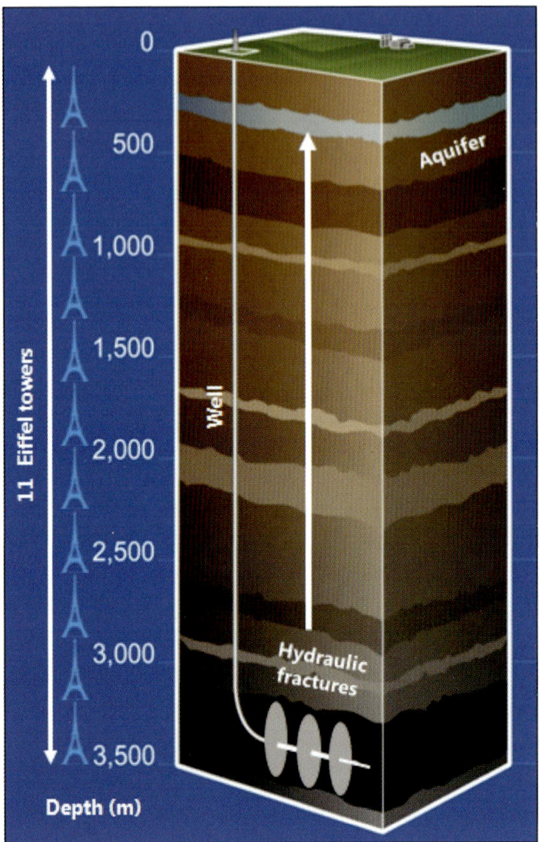

Figure 2 – As the maximum vertical propagation of a fracture is a few hundred meters, it is impossible to connect a shale oil and gas reservoir lying at a depth of 2,500 m with a surface aquifer.

…confirmed by microseismic monitoring data

Even though the seismic events generated by hydraulic fracturing[4] are no risk at all for stakeholders, they are an opportunity to create an accurate map of fracture growth in real time. The measurements therefore confirm the fact that the vertical penetration of hydraulic fractures can under no circumstances open a route between shale oil and gas formations and surface aquifers. After collecting thousands of microseismic data records,

4. See question 16.

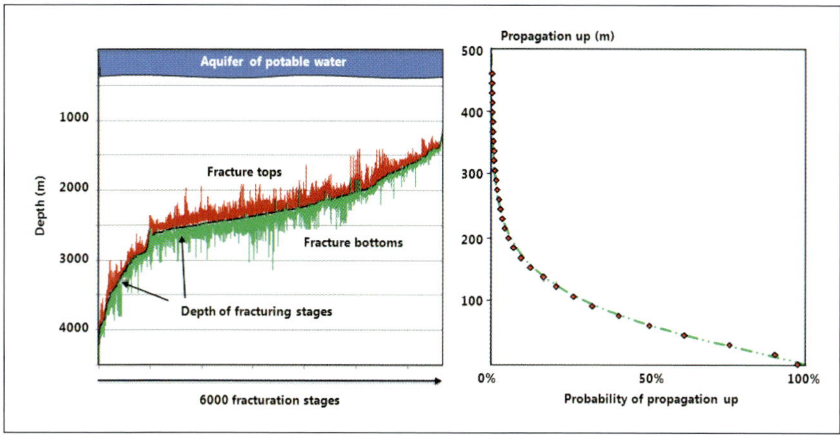

Figure 3 – Microseismic database containing over 6,000 fracturing stages. The "top" of the fractures shows that the vertical penetration does not exceed 350m and that consequently polluting a surface aquifer located much higher up is impossible (after Fisher and Warpinski).

Fisher and Warpinski[5] demonstrated on more than ten thousand fracturing stages that the vertical propagation of a fracture could not exceed 350m (**Figure 3**). So by taking a safety distance of approximately 750m between the zone to be fractured and the potable water-bearing layer, the probability of interaction between the two can be considered nil.

Physically nearly impossible

Even if the volume of fluid injected were to be increased infinitely, the fracture could not in most cases propagate right to the surface. At depths of 2,500-3,000m, where shale oil and gas is found, the horizontal thrust is almost always greater than the vertical thrust, thereby favoring the propagation of vertical fractures. At depths of 750-1000m however, the order is quite often inversed. As the vertical thrust becomes lower than the horizontal thrust, a fracture that reaches these depths often turns and propagates horizontally (**Figure 4**) preventing any connection with a potable water aquifer.

5. K. Fisher and N. Warpinski (2011), "Hydraulic Fracture-Height Growth: Real Data" SPE Annual Technical Conference and Exhibition Denver October 30–November 2, 2011. SPE 145949

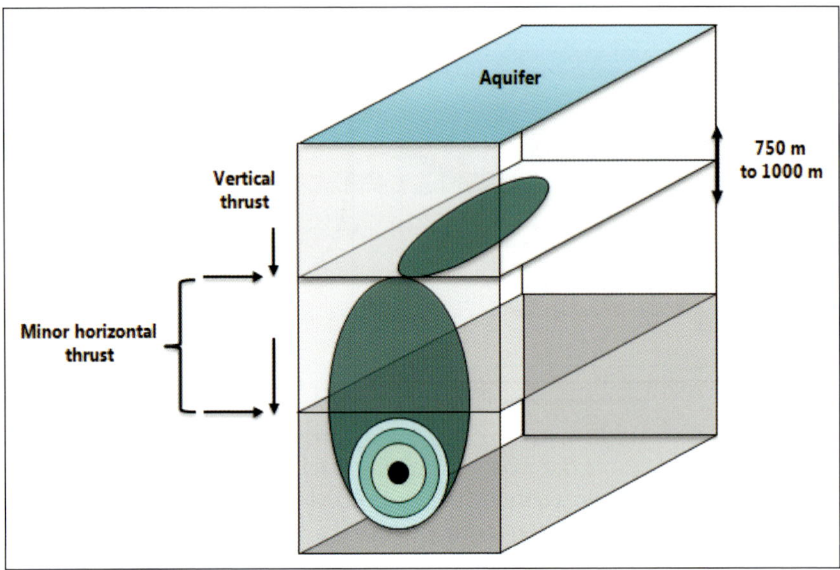

Figure 4 – At depth, the fracture propagates vertically since the minor horizontal thrust is lower than the vertical thrust. At 1,000m depth, the vertical thrust becomes lower than the minor horizontal thrust and the fracture turns and propagates horizontally. It is impossible for it to propagate to the surface and contaminate an aquifer.

It is therefore impossible for a hydraulic fracture, which begins at a depth of several thousand meters, to contaminate an aquifer. After several million hydraulic fractures made worldwide over more than 70 years, **NOT A SINGLE CASE** of groundwater pollution associated with such mechanism has ever been reported.

But the well is still "the weakest link"

A well is not just a tube that carries hydrocarbons to the surface; it is also an environmental barrier that protects potable water aquifers during drilling (drilling mud), hydraulic fracturing (fracturing fluid) and production operations (oil, gas and brackish water). In theory, the nesting of cemented casing[6] creates a seal that allows hydrocarbons to flow to the surface without ever coming into contact with the different formations

6. See Question 6.

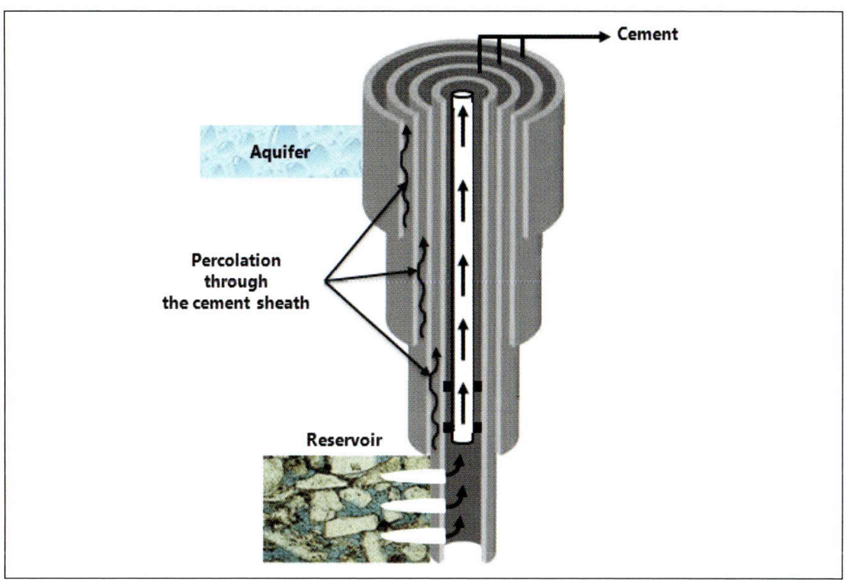

Figure 5 – The well acts as an environmental barrier which "channels" the fluid produced without it coming into contact with surface aquifers. Sometimes, owing to the aging of the cement, a small amount of fluid can percolate through, but cases of surface contamination are extremely rare.

they travel through, in particular the surface aquifers. This is referred to as *"well integrity"*.

Over time, the cement sheaths can become damaged and let small quantities of fluid (oil, gas or fracturing fluid) percolate through. But because of the nesting of the different casings a leak into the surface groundwater layers is highly unlikely. The construction of a well (and particularly the quality of the cementing operations) and its monitoring (constant measurement of pressure in the different cemented annular spaces) are the best precautions to guarantee well integrity. Nonetheless in the event of a leak, even a minor one, it is crucial to implement appropriate corrective measures, for example by perforating the casing at the leaking annular space to reinject cement at high pressure. Only one single case of a leak to the surface has been recorded to date in shale oil and gas production following a loss of well integrity (West Divide Creek – Colorado - **Figure 6**).

Figure 6 – The loss of well integrity in West Divide Creek[7] caused gas to travel back up to the surface. This is the only recorded example of surface contamination to date.

7. http://cogcc.state.co.us/Library/PiceanceBasin/WestDivide4_14_04summary.htm

Do shale oils and gases generate greenhouse gases?

Like any other industrial activity, the development and production of oil and gas emit GHG (greenhouse gases) in particular CO_2 and methane (natural gas - CH_4). Electricity generation, the gas flared and discharged into the atmosphere when wells are brought on stream and all the activities related to the transportation of equipment and personnel are the main sources of GHG emissions.

Compared with conventional fields, the extraction of shale oil and gas, which requires thousands of horizontal, multi-fractured wells, emits extra GHG. The main sources are electricity generation (more than half the emissions) and the gas flared and vented (over a third of the emissions). Drilling and fracturing activities contribute to less than 10% of emissions and fugitive emissions to 4%.

With emissions of between 25 and 40 $kgCO_2$/boe, the extraction of shale oil and gas emits 1.5 to 2 times more than the extraction of conventional oil and gas (20 $kgCO_2$/boe on average). Compared with LNG, which emits on average 60 $kgCO_2$/boe, they remain modest contributors.

These extraction emissions must really be put into perspective when compared with usage emissions. The thermal combustion of gas and coal emit respectively 300 $kgCO_2$/boe and 500 $kgCO_2$/boe and electricity generation 650 $kgCO_2$/boe for gas, and at least 1,300 $kgCO_2$/boe for coal. There is therefore a factor of 43 between the GHG emissions related to the extraction of shale oil and gas, and those related to the production of electricity using coal.

To reduce emissions generated by extraction, the oil industry has almost totally stopped flowing back wells in the open air, a practice which discharges part of the natural gas dissolved in the flowback effluent directly into the atmosphere. Instead green completions are used, in which the well is flowed back into a separator where the gas is recovered and sent directly via a pipeline to the treatment center. Other factors which can reduce emissions include using electricity from the national grid and transporting water in pipelines rather than tankers.

Lastly, even if the wellheads and the pipelines carrying the gas are not totally airtight, less than 1% on average of the natural gas produced is released into the atmosphere during the life cycle of a well. Cornell University estimated the GHG emissions at between 6-8%, demonstrating that the GHG footprint of the shale gas life cycle is thought to be higher than that of coal. The study is based on a sample of a handful of wells and harebrained hypotheses, in particular a CO_2 reference time of 20 rather than 100 years.

However, to control and minimize fugitive emissions, the oil industry is using surveillance stations equipped with infrared laser gas detectors more and more frequently.

Like any other industrial activity, oil and gas development and production activities are by nature emitters of GHG (greenhouse gases) in particular CO_2 and methane (CH_4). Flaring gas when the well is brought on stream, electricity generation and all the activities related to the transportation of people and equipment are the main sources of GHG emissions, referred to as extraction emissions.

Burning and electricity generation are the main culprits

Compared with conventional fields, the development of shale oil and gas, which requires thousands of horizontal multi-fractured wells, emits a larger quantity of GHG.

When PADs are developed, the emissions come from drilling rigs, the pumps used for hydraulic fracturing and also the numerous trucks required to transport the equipment, water, chemicals and sand to the site. However, as **Figure 1** shows, it is principally the methane which is flared and vented directly into the atmosphere during the well flowback phase, which is the main culprit. It represents more than four fifths of emissions whereas the drilling, fracturing and logistics activities represent just a fifth.

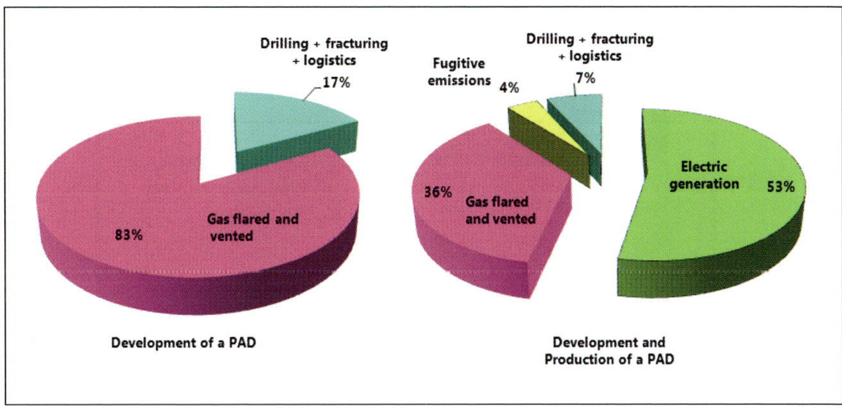

Figure 1 – During the PAD development phase, it is chiefly the gas flared and vented when the well is cleaned up which produces GHG. However, if we look at the entire cycle, electricity generation becomes the main culprit, responsible for over half the emissions for the entire cycle compared with a third for the flared and vented gas[1].

During the production phase, it is the generation of electricity required to run the installations (pumping and compression of effluents to the treatment center, fluid separation) which becomes the prime contributor. So over the lifetime of a field, electricity generation on average accounts for over half the extraction emissions and the flared and vented gas accounts for over a third. On the scale of a field lifetime, drilling and fracturing activities are low contributors, accountable for less than 10% of emissions.

Although the quantity of GHG generated can exceed 100 kgCO_2/boe during the initial development phase (intensive drilling and fracturing activity, gas flared and vented when the well is flowed back - **Figure 2**), the volume rapidly drops to stabilize at values of between 25-40 kgCO_2/boe.

1. JA Costa (2011), "Estimation and comparison of the Greenhouse Gas Emissions between a shale gas project and a conventional gas project", Total Internship report.

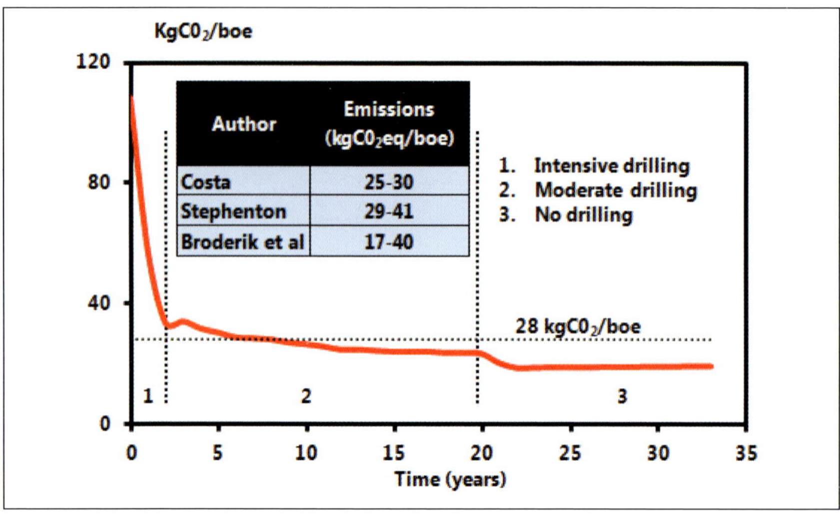

Figure 2 – The very high GHG emissions rate at the start of development subsequently plummets[2]. On average, the quantity of GHG emitted by a shale oil and gas development is somewhere between 25 and 40 kgCO_2/boe[3,4,5,6].

The difference compared with a conventional development is minimal

Due to the fact that associated gas is no longer flared on new oil fields, the oil industry has made enormous progress, reducing the extraction emissions on average from 100 kgCO_2/boe[7] to 20 kgCO_2/boe - a rate

2. JA Costa (2011), "Estimation and comparison of the Greenhouse Gas Emissions between a shale gas project and a conventional gas project", Total Internship report.
3. T. Stephenson (2012), "Modeling the relative GHG emissions of conventional and shale gas production", IPIECA workshop 6-7 march 2012.
4. J. Broderick, K. Anderson, R. Wood, P. Gilbert, M. Sharmina, A. Footitt, S. Glynn, F. Nicholls (2012), "Shale gas: an updated assessment of environmental and climate change impacts", Tyndall Centre University of Manchester.
5. M. Jiang,W.M. Griffin, C. Hendrickson, P. Jaramillo, J. VanBriesen and A. Venkatesh (2011), "Life cycle greenhouse gas emissions of Marcellus shale gas", IOP PUBLISHING ENVIRONMENTAL RESEARCH LETTERS Environ. Res. Lett. 6 (2011), 034014 (9pp).
6. D.J.C. MacKay and T.J. Jones (2013), "Potential greenhouse gas emissions associated with shale gas extraction and use", UK Department Energy and Climate Changes.
7. The figure 100 kgCO_2/boe remains the average for many mature fields in equatorial Africa and in Russia.

equivalent to the average rate usually observed on conventional gas fields. With extraction emissions somewhere between 25 and 40 kgCO_2/boe, shale oil and gas are very reasonable contributors, in particular when compared to Liquefied Natural Gas, a strong contributor process which emits on average 60 kgCO_2/boe. The reduction of these emissions relies essentially on limiting the quantity of gas flared and vented during the lifetime of a field. Remember that these **extraction emissions**, concerning the development process must be put firmly into perspective when compared with the **usage emissions** resulting from the combustion of fossil fuels used either for heating or to generate electricity (**Figure 4**). So the thermal combustion of gas emits 300 kgCO_2/boe and that of coal 500 kgCO_2/boe. Regarding electricity generation, the gas cycle emits on average 650 kgCO_2/boe whereas for the coal cycle, it is at least 1,300 kgCO_2/boe. There is therefore a factor of 43 between the GHG emissions related to shale oil and gas production and those related to the production of electricity using coal.

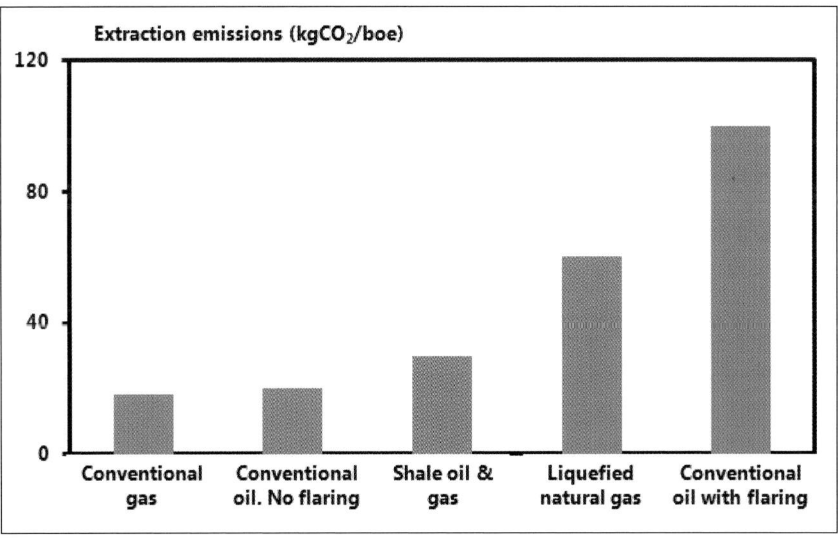

Figure 3 – Comparison of extraction GHG emissions corresponding to different types of developments. Although shale oil and gas developments emit more than conventional gas developments, the quantity is much lower than that emitted by the LNG production process.

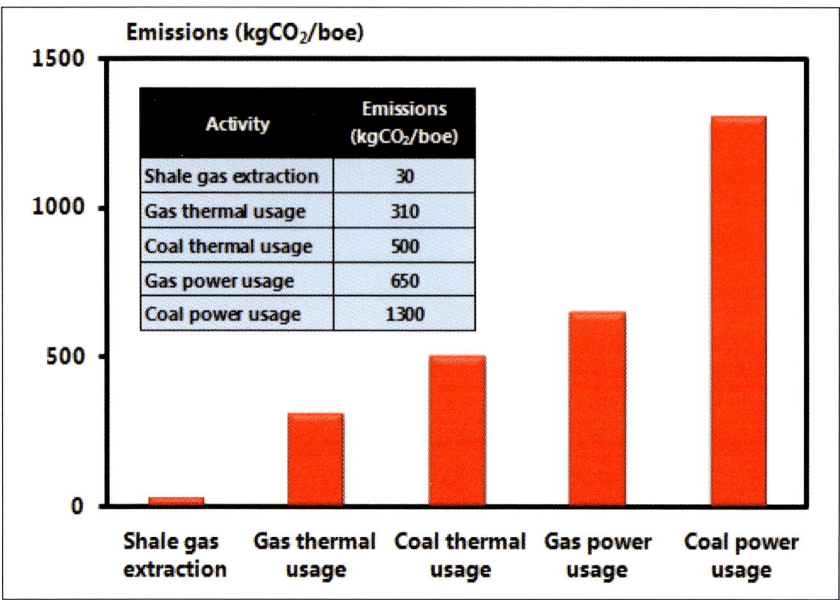

Activity	Emissions ($kgCO_2$/boe)
Shale gas extraction	30
Gas thermal usage	310
Coal thermal usage	500
Gas power usage	650
Coal power usage	1300

Figure 4 – Although the techniques for developing shale oil and gas emit more than those used for conventional gas, the extraction emissions are marginal when compared with the usage emissions resulting from combustion or electricity generation, in particular coal-fired electricity production[8].

Reducing vented gas by using green completions and tanker transportation

The difference in extraction emissions between the upper value (40 $kgCO_2$/boe) and the lower value (25 $kgCO_2$/boe) can be explained essentially by the well flow back method (**Figure 2**), a phase during which significant quantities of natural gas, whose GHG potential is 25 times higher than that of CO_2, are discharged into the atmosphere. To try and reduce these emissions, the oil and gas industry continues to innovate. The flagship technology involves limiting the venting of natural gas in the well flowback phase by using "green" completions. The old method entailed flowing back the well in the open air and consequently releasing

8. D.J.C. MacKay and T.J. Jones (2013), "Potential greenhouse gas emissions associated with shale gas extraction and use", UK Department Energy and Climate Changes.

Figure 5 – Green completion involves a separator – a large horizontal tank used to separate the gas from the water recovered during the flowback phase thereby avoiding any fugitive emissions inherent to well flowing back in the open air.

part of the natural gas dissolved in the recovered water into the atmosphere. A green completion involves cleaning up the well into a separator (a large tank, often installed horizontally - **Figure 5**) in which the gas is separated from the water and sent directly by pipeline to the treatment center. For just a small extra cost, green completions not only avoid the release of part of the dissolved gas into the atmosphere, but they also go a long way to avoiding gas flaring during the first days of production. The results in terms of reducing extraction emissions are substantial[9].

Other factors which can reduce emissions include using electricity from the national grid rather than diesel to power the motors and pumps on site and using pipelines instead of tankers for the transportation[10] of the water, chemicals and sand required for hydraulic fracturing.

The controversial question of fugitive emissions

During the field production phase (conventional or not), the wellheads and pipelines carrying the gas to the treatment center and then to the

9. S. Palttsev & V. Karplus (2013), "Fugitive methane from shale gas production less than previously thought", Global Change Spring 2013.

10. See Question 14.

market are not totally airtight. On average, less than 1% (0.5% according to a recent study on 190 fields in the US[11]) of the natural gas produced is released into the atmosphere during the lifecycle of a well, which represents approximately 4% of the total emissions (**Figure 1**). These are called *"fugitive emissions"*.

By estimating that fugitive emissions represented on average not 1%, but 6-8% of the gas produced, Cornell University[12] arrived at a startling conclusion: the GHG footprint of the shale gas life cycle is estimated to be higher than that of coal, a strange assertion when we look at the figures presented in **Figure 4**. Just as for the *Gasland* film, this study, based on a handful (five) of unrepresentative wells, and harebrained hypotheses, caused a general uprising against shale oil and gas. The study overlooks the fact that gas-powered electricity generation is twice as efficient as coal-powered generation and, above all, the study used a reference time of 20 years rather than 100[13], thereby quadrupling the GHG potential of fugitive natural gas compared with the CO_2 emitted by the coal cycle.

Whereas standard fugitive emissions (1% with a reference time of 100 years) are in line with extraction emissions of between 25 and 40 kgCO_2/boe (**Figures 3** and **4**), the unrealistic hypotheses put forward by Cornell University produce unrealistic figures of 1,200 kgCO_2/boe for 6% and 1,850 kgCO_2/boe for 8% fugitive emissions. By forcing the emission figures and the reference time in this way the gas life cycle effectively becomes a larger contributor to GHG emissions than the coal life cycle.

In an effort to control and minimize fugitive emissions, the oil industry is using monitoring stations equipped with infrared laser gas detectors more and more frequently. They can detect very low concentrations of natural gas in air in just a few seconds, and rapidly correct any natural gas leak that cannot be attributed to standard fugitive emissions.

11. Allen T *et al.* (2013), "Measurements of methane emissions at natural gas production sites in the United States", http://www.pnas.org/content/early/2013/09/10/1304880110.abstract.

12. Howarth *et al.* (2011), "Methane and the greenhouse-gas footprint of natural gas from shale formations".

13. See Appendix 11 for further information.

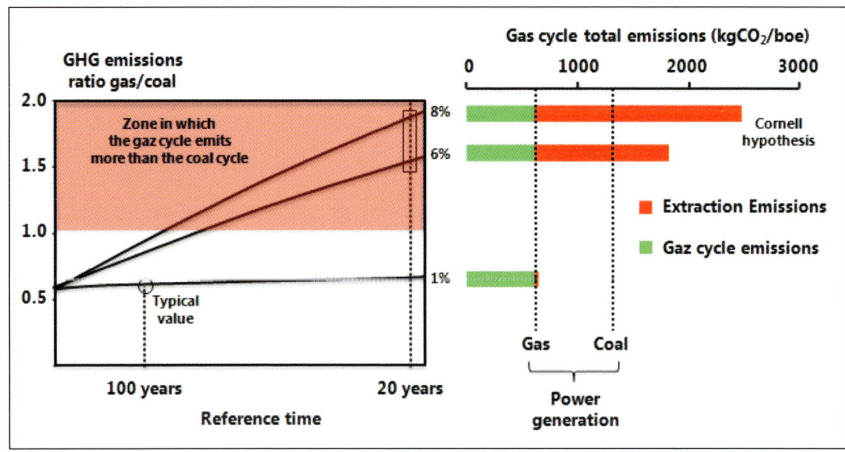

Figure 6 – Fugitive emissions which represent a maximum of 1% give a global footprint for the gas cycle of around 650 kgCO_2/boe. The hypotheses put forward by Cornell University would triple (for 6%) and quadruple (for 8%) the emissions values.

The shale oil and gas revolution, a durable project or an economic bubble?

An economic bubble is the result of a situation which appears favorable and sustainable for a certain period of time but which in the long-term proves to be the result of an over-confident illusion. Fed by positive rumors, myths, euphoria and collective imitation, it suddenly bursts when reality catches up. The scale, suddenness and the euphoria generated by the shale oil and gas revolution in the United States causes fear among certain people that this, too, might prove to be a speculative bubble.

This euphoria effectively moved many independent US companies to go into debt and produce gas exclusively. The crash in gas prices in 2012 caused these same independent companies to sell many assets at extremely low prices to repay their debts. Since then most of them have produced negative cash flows.

In addition to the contextual crash in gas prices, certain experts consider that shale oil and gas production and ultimate resources in the United States are wildly overestimated. Standing somewhere between 30 and 40% per year, the decline in well production means that new wells must be constantly added, causing an alarming increase in the cost of replacing the production wells that already exist. This kind of inflation would not be viable in the medium-term and would lead to an economic bubble, causing the withdrawal of investors, the shutdown of developments and a rapid marked decline in production. Prices would then soar and the world gas market would be thrown off balance.

For several reasons, this disaster scenario cannot happen in the United States in the medium term:

(1) The well-by-well analysis is deceptive. North America has a vast portfolio of wells which acts as a buffer and which, despite the slow-down in investments, staves off global decline.

(2) Production decline in the Barnett and Haynesville shales, where the break-even points are too close to the current gas prices, has been more than offset by the spectacular progression in the production of the Marcellus field, which is much more profitable.

(3) The long term stability of oil prices enabled independent companies who have moved a high percentage of their activity from dry gas to wet gas and oil, to repay their debts.

(4) The export of LNG to Europe or Asia by 2017 should smooth out gas prices.

However, although the production of shale gases in the United States is an activity that is cushioned enough to exclude the possibility of an economic bubble, its sustainability is not guaranteed. Shale gases are not easy prey and producing them sustainably and economically requires an in-depth review of the applicability of the trial and error method, the limitations of which are now only too apparent. These negative experiences encourage operators to implement quality models that are more scientific and boost the recoverable resources per well.

An economic bubble[1] is the result of a situation which, rightly or wrongly appears favorable and sustainable for a given period of time but which in the long term, proves to be the result of an over-confident illusion.

To begin with, it is fed by positive rumors, myths, euphoria and collective imitation. Seeing others win makes other people want to *"join the party"*. The belief of speculators then triggers a sudden, excessive increase in the value of a given market (exchange, real estate, raw materials, etc.) relative to its intrinsic value.

At a later stage in the cold light of day, the loss of confidence causes an immediate turnaround and the market is exposed to an instant crash, just like a bubble that floats upwards and then bursts. Euphoria gives way to panic and real bankruptcy takes hold, causing disastrous socio-economic collateral damage.

1. http://en.wikipedia.org/wiki/Economic_bubble

Although the history of mankind has been marked by economic bubbles at regular intervals[2], two significant events occurred one after the other in the last 15 years: the internet bubble in 2001 and the subprime crisis in 2008. The latter was a consequence of real estate speculation where the individual bankruptcy of borrowers given access to virtually interest free credit, was to be covered by the illusion of resale at a higher price. It placed everyone in this virtual rising market at the mercy of a turnaround. We all know what happened next.

Over the last five years, the shale oil and gas revolution has generated an almost exponential growth in oil and gas production in the United States. The scale, suddenness and unprecedented general euphoria it induced have caused fear among certain people that this is just another economic bubble[3]. Alongside environmental risks, this argument is often taken up by opponents once they have run out of technical arguments.

From a purely contextual situation to an "oil & gas" bubble?

The euphoria in the period from 2009 to 2012 caused many independent US companies to go into debt to board the shale oil and gas train. Dedicated investments increased exponentially from 5 GUS$ in 2006 to over 80 GUS$ in 2013[4]. The crash in gas prices in 2012[5] in a market that was not prepared to withstand such a massive increase in production[6] led these same independent companies[7] to sell off numerous assets at low prices to repay their debt. A recent analysis of the financial performance[8] of 35 independent companies which account for 40% of the overall shale oil and gas production in the United States (i.e. over 3 Mboe/day) shows

2. Tulip mania in XVII[e] century in the Netherlands; an economic crash related to the speculation on shares of the Southern Seas Company in 1720; massive real estate speculation in Paris, Berlin and Vienna at the beginning of the 1870s, which ended in the major stock market crash in Vienna 1873.
3. http://www.huffingtonpost.com/steve-horn/reports-shale-gas-bubble-_b_2718354.html
4. Oswald Clint (2013), "The dark side of the golden age of shale gas and tight oil".
5. See Question 3.
6. In 2011 production was 4 times greater than demand.
7. Fôret sold 1.3 G$ worth of assets after announcing disappointing results.
8. I. Sandrea (2014), "US shale gas and tight oil industry performances : challenges and opportunities", Oxford Energy Comment.

that over the last five years their financial situation has gradually deteriorated. In spite of having moved a high percentage of their activity from gas to oil, the analysis shows that their income barely covers their investments. Almost all of them displayed negative cash flows proportional to the repayment of their debt. Although independent companies were the first to suffer, certain majors, including Shell[9], have also acknowledged having to divest some of their assets to arrest a financial hemorrhage caused by disappointing results. Exxon Mobil however, considered its heavy financial losses as purely contextual.

In addition to the contextual crash in gas prices owing to unexpected overproduction, certain experts are wary of the promises being made. They consider that future production levels and the ultimate reserves are wildly overestimated[10]. Remember that even if there are significant quantities of hydrocarbons in place in the source rock, just a small percentage is recovered and a minor error in the recovery process could lead to major errors regarding resources. In 2011, the EIA estimated that of the 1,200Gboe of shale gas in the United States, 11% (i.e. 130 Gboe) would be technically recoverable. In 2012, the estimate of 11% was deemed too optimistic and changed to 8%, i.e. an absolute reduction of 30Gboe in technical resources. This marked decrease is a result of well behavior in which production rapidly declined; a phenomenon inherent to the low permeability of the source rock and to the recovery technique used. While the annual production on conventional oil fields declines at a rate of between 3 and 6% per year and 15-20% per year for gas fields, where shale oil and gas is concerned, the rate of decline is somewhere between 30 and 40% per year[11]. In 2001 when the United States was producing only conventional gas, the rate of decline was around 23% per year. In 2012 the figure topped 33%. To maintain the current US production gas rate of 11 Mboe/day, 3.7 boe/day need to be replaced every year. The increasing cost of this replacement is cause for concern - for example, that of Haynesville is said to have increased by 40% between early 2008 and mid-2010[12].

9. Bloomberg article dated May, 2014.

10. http://shalebubble.org/wp-content/uploads/2013/02/SWS-report-FINAL.pdf

11. S. Farell (2013), "US oil production : impact of high unconventional decline rates", PCF Energy.

12. A.E. Berman (2012), "After the gold rush: a perspective of future US, natural Gas Supply and Prices", ASPO conference 2012 Vienna Austria.

According to critics, the trial and error method[13] which entails drilling and fracturing new wells one after the other to offset the extremely rapid decline of existing wells is not viable in the medium-term and will generate an economic bubble. The withdrawal of investors would stop developments and trigger a rapid marked decline in production.

Next, prices would soar and the United States would lose their competitive edge in the refining, petrochemicals and manufacturing industries. The drop in production would dash any hopes of becoming exporters as of 2017, making all the investments committed for the transformation of LNG terminals null and void. The world gas market would be thrown off balance, particularly in Europe where the weight of long-standing suppliers (Russia, Norway) would be reinforced. This disaster scenario predicted by certain experts in 2012, cannot happen in the medium-term for a number of reasons:

- The well-by-well analysis is deceptive in view of the amazing number of wells drilled, fractured and brought on stream over the last few years. At the end of 2013, North America (United States + Canada) had a portfolio of some 110,000 shale gas wells on fields at different stages of maturity. In spite of the rapid individual decline, this considerable stock acts as a buffer which helps limit the overall decline for several years without having to continue drilling and fracking frenetically. The significant slow-down in investments (i.e. drilling and fracking activities) which began in 2009 on the Barnett field did not lead to plummeting production figures – far from it (**Figure 1**).
- The low decline in production on the Barnett and Haynesville developments, where the break-even points[14] are too close to current gas prices (**Figure 2**), were more than offset by the spectacular progress of production on the much more profitable Marcellus play. This is the result of less restrictive exploitation conditions (shallower formation) but also of higher reserves per well. The Marcellus field also produces more liquid (oil and condensates) which, before the recent drop of the oil prices, significantly reduces its break-even point compared with Barnett and particularly Haynesville that

13. See Question 11.

14. Gas price which cancels out the NPV10%.

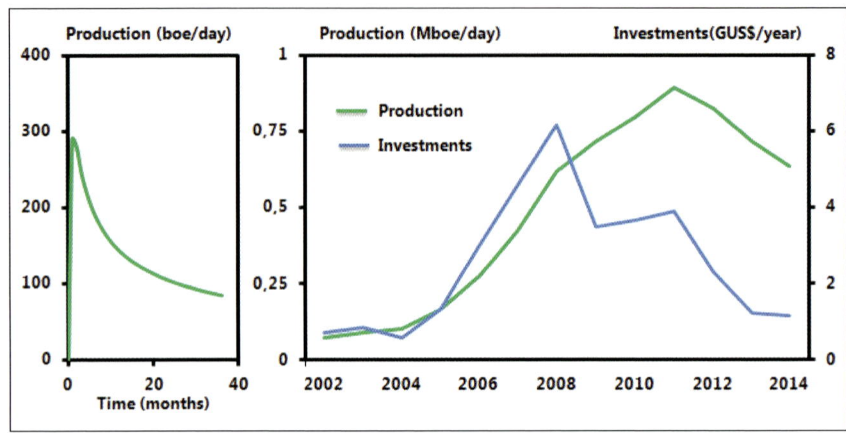

Figure 1 – The decline of the Barnett field (right) does not reflect the individual decline of a well (left). A latency of several years can be seen between the slowing down of investments which began in 2009 and the decline in production which started in 2012. (Source: Wood McKenzie)

produce essentially dry gas. However, they bear no relation to the "*real*" shale oils such as those of the Bakken field where the break-even point is in the range of US$ 60/boe (**Figure 2**). Development of the Marcellus field should gather speed in the next few years, and IHS CERA[15] anticipates that its global production (oil+gas), which will exceed 2Mboe/day in 2014, could conceivably reach 5Mboe/day by 2035.

- Far from being over[16], the gas price adjustment cycle is estimated to continue for roughly another five years, which will continue to make shale gas producers "*suffer*" periodically. The year 2017 which should herald the start of LNG exports to Europe or Asia, will smooth out gas prices. Investors therefore anticipate that the financial performance of gas producers is set to remain fairly poor for the next five years, before gradually picking up again. However, it is unrealistic to think that the price of gas in the United States can remain under US$ 30/boe in the long term. The smoothing out of gas prices will cause the US petrochemical and manufacturing industries to lose some of their competitive edge which currently is unusually high.

15. IHS CERA (2014), "The evolving long-term outlook for North American natural gas".

16. The price of gas, which in 2012 was under US$ 11/boe increased to US$ 30 during the very harsh winter of 2013. It had fallen back down to US$ 23 by summer 2014.

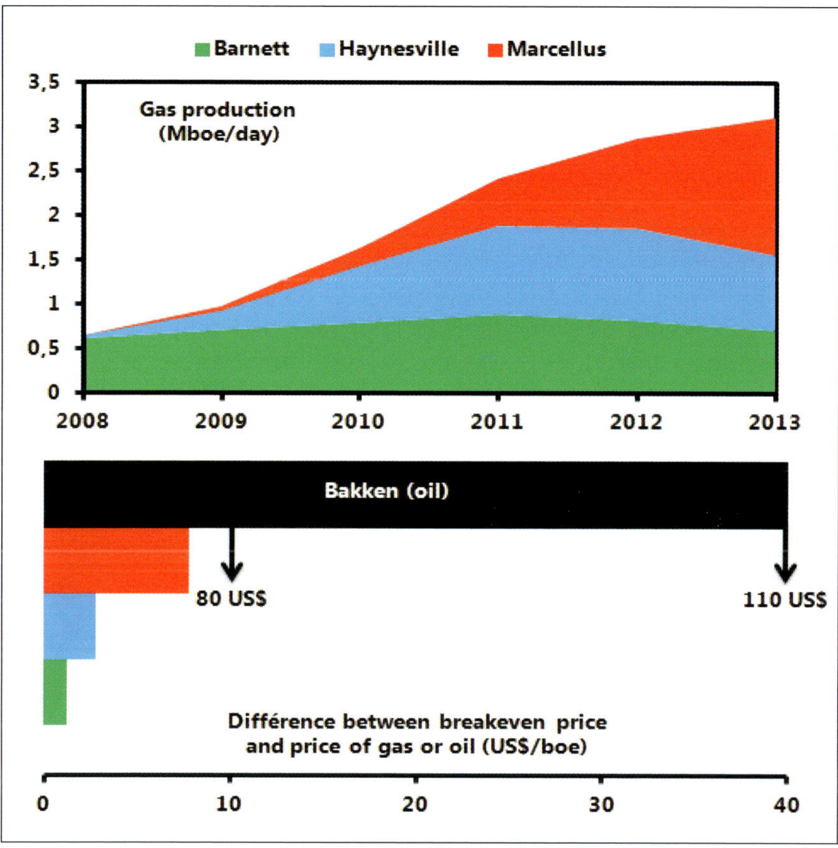

Figure 2 – The production decline on the Barnett and Haynesville developments where the break even points are too close to current gas prices, has been offset by the spectacular progress of the much more profitable Marcellus development. A very different scenario from that of Bakken (shale oil) where the break-even point is much lower than current oil prices (source: Wood Mc Kenzie)

Similarly but 2 years later, the oil market which has seen the arrival of unexpected barrels recently experienced a very sharp decrease in prices dropping from US$ 110/bbl in June 2014 to less than US$ 70/bbl beginning December 2014. Small independents but also major companies, victims of their operational success, will therefore suffer twice as much from low oil & gas prices. However as a world spot market, oil is much more sensitive (positively if quotas are imposed) than gas to geopolitical events.

However shale oil & gas production in the United States is an activity that is amortized enough to exclude the possibility of an economic bubble. Large capitalistic arctic or deep water oil & gas projects should suffer much more from the instability of the oil prices than shale oil which can quite easily adapt to a stop and go strategy. Nonetheless, its sustainability is not guaranteed owing to unknowns regarding the economic life of wells and long term uncertainties regarding prices. Shale gas is clearly not an easy prey, and producing it sustainably and economically requires an in-depth review, even in the United States, of the applicability of the trial and error method, the limitations of which are now only too apparent. The low prices should also delay the export of the US revolution which in many places is not currently economic.

These negative experiences encourage operators to adopt more scientific approaches, in particular those that help pinpoint sweet spots[17] more accurately, in order to increase both the initial production and ultimate recovery of wells, while limiting their number. Thanks to such technologies for example, the break-even price for gas was reduced by 35% on the Marcellus between July 2012 and early 2014.

17. R. Weijermars (2012), "Assessing the economic margins of sweet spots in shale gas plays", 2012 EAGE First break.

What are the opportunities and risks of developing shale oil and gas in Europe?

Soaring oil and gas prices between 2000 and 2010 significantly increased Europe's energy bill. This contributed extensively to increasing sovereign debts, reducing investment, boosting production costs, slowing down consumption and ultimately to the loss of many jobs. The figures show that for the same social system, the French national debt is the sum of the cost of energy and the repayment of debt. During the golden sixties, developed countries had built their growth strategies on the basis of energy that cost next to nothing.

Therefore, if Europe wants to engage in sustainable growth again, it must reduce its energy bill and to do this, a calm, responsible, coordinated debate based on objective data needs to be had. The question of whether to develop (or not develop) shale oil and gas in Europe is based on a sizeable unknown – the actual state of resources.

According to the American Energy Agency, Europe is thought to hold approximately 93Gboe, 85% of which is gas. To develop them, two 2020/2050 scenarios have been put forward: some shale gas (gas plateau corresponding to twice the national consumption of France) and the shale gas boom (plateau corresponding to four times the national consumption of France). To achieve these production plateaux, somewhere between 23,000 and 50,000 wells would need to be drilled and 100 to 300 drilling rigs mobilized. The footprint and water usage would be respectively between 250 km^2 and 500 km^2 (equivalent to Lake Geneva) and between 500 million and 1 billion m^3. These figures may seem high but are actually marginal on a European scale. By 2035, Europe's energy dependency which would reach 95% without any shale gas development would be reduced to somewhere between 78% and 62% and the price of gas by 10% to 20%. The proposed scenarios would provide an average annual growth of between 0.3 and 0.6% and create between 600,000 and 1 million jobs. To carry this major project through to fruition, three barriers must be overcome.

(1) The first is geological. Are European source rocks, better, the same or worse than their US counterparts? The only way to answer this question is to drill exploration wells.

(2) The second is economic. The viability of this kind of project will depend on our capacity to drill and fracture wells at acceptable costs. Unless truly exceptional reservoirs are found, the European market as it stands would not be conducive to the cost-effective production of European shale gas. The emergence of a local competitive oil services market first requires determined political backing.

(3) The third covers all the environmental, societal and cultural issues. In a densely-populated urbanized Europe, hydraulic fracturing, water supply, microseisms and surface impact are seen as a series of threats and to change this perception will take more than scientific arguments. Winning over public opinion requires pedagogy, transparency, appropriate communication campaigns and commitment regarding local communities, with a review of certain regulations. The game must be win/win. Mining law, which is essentially favorable to the national governments of all Member States, is obviously not a factor that will encourage local authorities and local communities to be proactive.

The crushing weight of the European energy bill

Europe's baby boomers had a double stroke of luck: they were born into a growth society with full employment, and they lived through a time of peace. Although the European Union has many failings, at least it has the virtue of protected the old continent from any conflict for almost 70 years, a first in History, because for fifteen centuries, the Europe of Nations had ripped itself apart, leaving in its wake violence, destruction and desolation.

However, this generous European Union, progressively consolidated in economic and then monetary terms, also suffers from deep-seated divides in which energy plays a pivotal role.

Following the explosion in oil prices, which rose from US$ 30 in 2000 to over US$ 100 in 2010, Europe's energy bill has ballooned. Coupled with the subprime crisis, it has contributed significantly to causing sovereign debt to soar, reducing investment, increasing production costs and

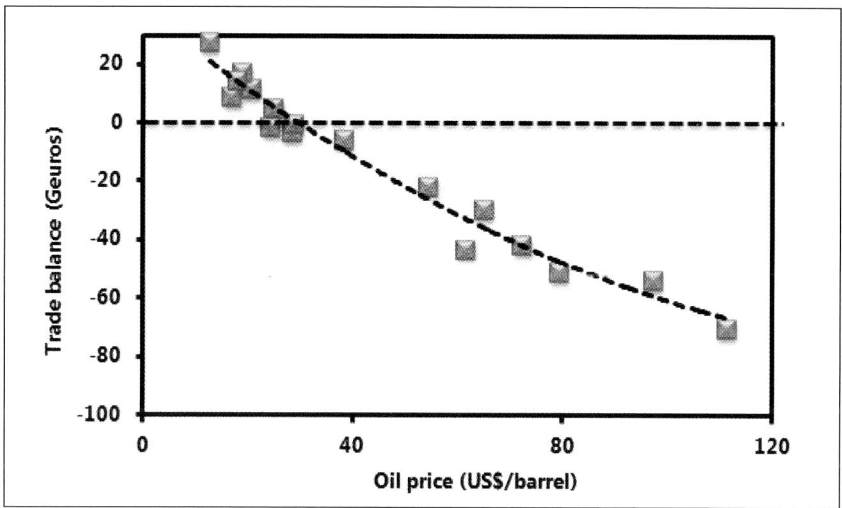

Figure 1 – The French trade balance mirrors oil prices perfectly[1].

slowing down consumption. In 2011 and 2012, oil prices alone are estimated to have stunted European growth by between 0.6% and 0.8%[2]. For example over the last decade France has seen its energy put a strain on its trade balance[3] where the deficit mirrors oil prices perfectly (**Figure 1**). In 2012, the oil and gas bill represented almost 90% of the French trade deficit.

The impact of the oil and gas bill on the French economy is nothing new and, not surprisingly, it dates back to 1973, when the first oil crisis occurred. From this date onward, growth began to dwindle, sovereign debt appeared and the French people experienced mass unemployment. A comparison of the increase in GDP (850 Geuros of cumulated growth between 1993 and 2010), the increase in debt (1,000 GEuros) and the cost of oil and gas imports (750 Geuros) clearly shows that **for the same social system**, the French national debt (and also that of most European countries) is **the sum of the cost of energy and the repayment of debt** (**Figure 2**). During the golden sixties, developed countries had built their growth strategies (and consequently, their social system) on the basis

1. Source: Customs (estimated FAB data).
2. http://www.insee.fr/fr/ffc/iana/iana7/iana7.pdf
3. http://lekiosque.finances.gouv.fr/Appchiffre/Etudes/Thematiques/A2012.pdf

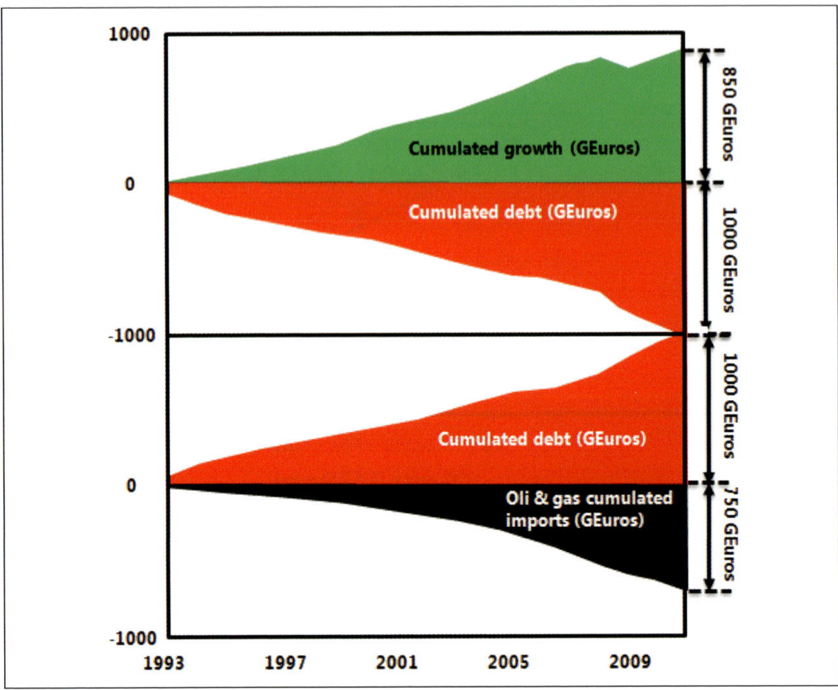

Figure 2 – Growth, debt and oil and gas imports[4]. For the same social system, French national debt can be considered as the sum of the cost of hydrocarbons and the repayment of its debt.

of energy that cost next to nothing. The first oil crisis made short work of this privileged position and forced the same countries to put themselves in debt to import oil and gas at high cost while continuing to finance social systems that were well beyond their means – a hellish downward spiral into which the Europeans were inevitably drawn.

... coupled with an energy divide

Yet the European countries, on the pretext of an age-old inheritance or national political decisions, continue to build their own strategies without any real desire to consolidate a common energy policy (**Figure 3**). What with the rejection of nuclear energy, the comeback of coal and a possible entry into shale oil and gas, environmental concerns and promises of

4. Sources: Energy Funds Advisors.

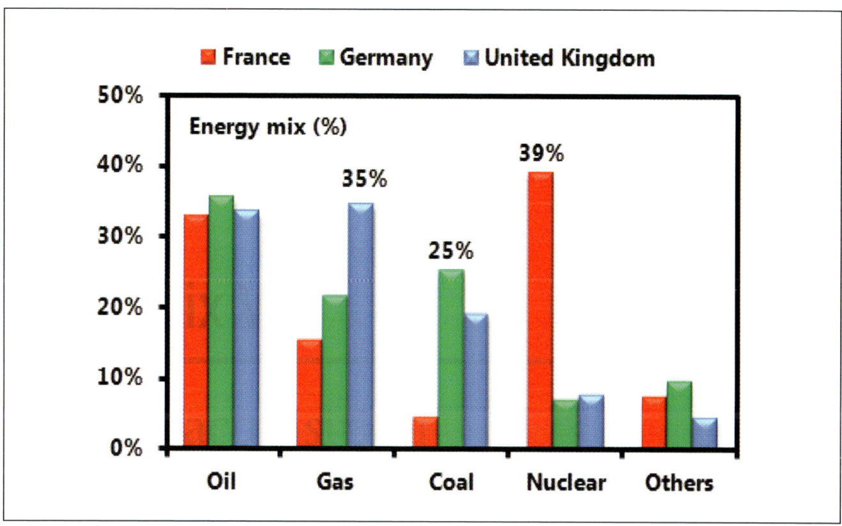

Figure 3 – The energy mixes of a nuclear France, a coal-fired Germany and a gas-fired United Kingdom[5] are the reflection of European divides regarding electricity generation.

low-cost energy, things are going from the sublime to the ridiculous. While Germany is withdrawing from nuclear energy but has begun mass importation of coal (+12,5% between 2009 and 2013) and France has shelved the shale gas file, even excluding any exploratory phase, the British government is creating *"the most favorable tax regime in the world to encourage the development of unconventional hydrocarbons in Great Britain"*[6].

If Europe wants to engage in sustainable growth again, it must reduce its energy bill, and this objective is not really compatible with discordant energy policies which send out contradictory messages to public opinion and to industry as a whole. What Europe needs is a calm, responsible, coordinated debate based on objective data, as the question of whether to develop (or not develop) shale oil and gas in Europe is based on a sizeable unknown – the actual state of resources.

5. BP 2102 outlook

6. Le Figaro 14/08/2013 INTERNATIONAL, "Cameron prend position pour les gaz de schistes", (Cameron takes a stand pro-shale gas).

Yet, to make the dream of a great European project a reality…

According to the latest evaluations from the EIA[7], Europe is thought to hold 93 Gboe, 85% of which is shale gas and just 15% shale oil. The main resources are most likely in Poland and in France (28 Gboe each) and to a lesser extent in Romania, Denmark, Holland and England. However, these estimates which are based on simplistic volume calculations must be considered with great caution. Only data collected during the drilling of exploration wells will actually give a clear picture of the real stakes[8]. Assuming that these hypotheses were to be confirmed at least in part, to what extent could these resources reduce Europe's energy dependency on its main gas suppliers (Norway, Russia and Algeria)? Would they trigger significant positive impacts on growth, energy prices and employment?

The gas dependency of Europe was in 2013 equal to 67%. Its cost was nearly 100 GEuros that is 0,7% of its global GDP. Without any shale gas development and given the decline of the conventional gas fields in the North Sea, the dependency at the horizon 2035 will reach 95%.

After applying a number of technical (non-recoverable resources), environmental & societal (national parks, densely populated urban regions) and economic (excessive development costs) filters to EIA resources, two European scenarios spanning the period between 2020 and 2050 (**Figure 4**) have been put forward[9]. The first, some shale gas, ambitiously forecasts a peak at 1.2 Mboe/day[10] to produce 8.6 Gboe (11% of EIA gas resources) by 2050. The second, shale gas boom predicts a pseudo plateau of 2.7 Mboe/day with resources of 20 Gboe by 2050 (25% of resources). In the absence of any real production data, an average equivalent Barnett well (550,000 boe of ultimate resources) has been considered. To put the stakes in context, the first scenario would cover just short of twice the national consumption of France and the second, four times the national consumption.

---

7. See Question 2, Figure 5.
8. Thomas Moore Institute (2012), "Gaz de schiste et Europe. Analyse comparative dans 14 pays Européens", Note de Benchmarking 14.
9. J. Williams, M. Winter, P. Summerton & S. Billington (2013), "Macroeconomic effects of european shale gas production" POYRY – Phase I: gas and power market impacts - Phase II: Employment impact". http://www.poyry.co.uk/sites/poyry.co.uk/files/public_report_ogp__v5_0.pdf, Novembre 2013.
10. Remember that the daily consumption of France is 4 bcf/day.

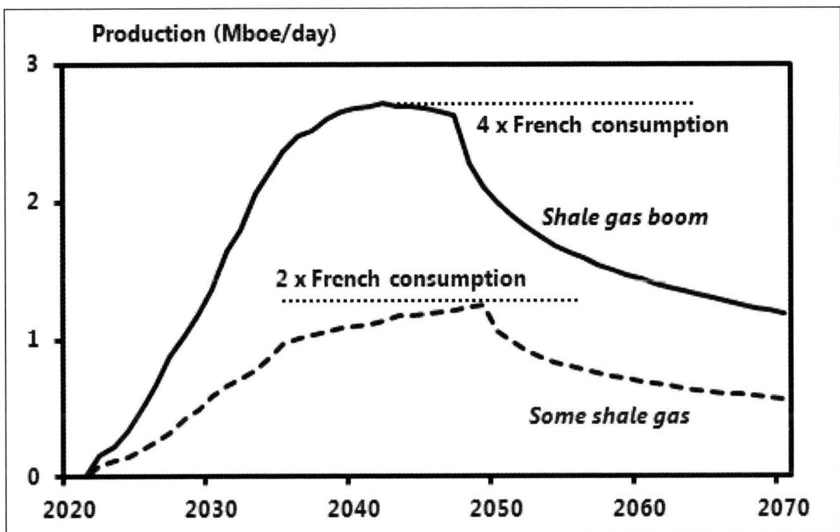

Figure 4 – Two shale gas development projects.

To create the production graphs shown in **Figure 4**, approximately 23,000 wells would need to be drilled for the some shale gas scenario and over 50,000 in the shale gas boom scenario, with peaks of 1,000 and 3,000 wells respectively around 2035 (**Figure 5**) which, assuming PADs of 10 wells, would require the mobilization of 100 to 300 drilling rigs.

In terms of the footprint over the 28 countries of Europe, the total surface area used would be somewhere between 250 km^2 and 500 km^2, equivalent to that of Lake Geneva[11]. Remember that to produce the same quantity of energy using solar panels or wind farms, the required surface area would be 10 times greater[12].

To satisfy the needs of hydraulic fracturing, the total water usage would be somewhere between 500 million m^3 and 1 billon m^3 over thirty years. These figures may seem high but are actually marginal on a European scale. So in the shale gas boom scenario, the peak water usage would be around 60 million m^3 in 2035, a value to be compared with annual water consumption in France which, for 2012 alone was 33 billion m^3. These

11. The surface area of Lake Geneva is 581 km^2.

12. See question 13, Figure 3.

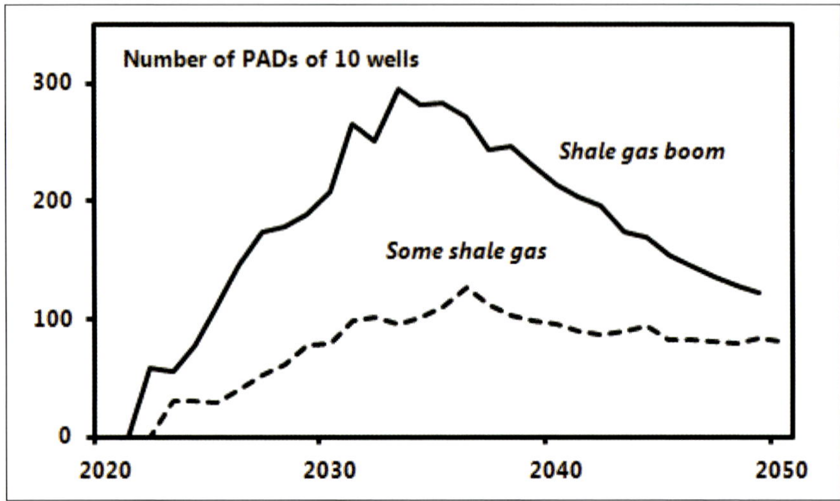

Figure 5 – Number of ten-well PADs required to produce the two scenarios. In the case of some shale gas, approximately 23,000 wells need to be drilled whereas the shale gas boom would require over 50,000 wells between 2020 and 2050.

results confirm that the nuisances in terms of footprint and water usage are purely local.

The most significant macroeconomic impacts resulting from these two scenarios would be[13].

- By 2035, Europe's energy dependency on its main gas suppliers (Russia, Norway, Algeria) would be reduced to 78% in the case of some shale gas and to 62% in the case of the shale gas boom (compared with 90% in the absence of any development).
- By 2035, the development of shale gas would reduce the price of gas by 20% in the case of the shale gas boom and by 10% in the case of some shale gas.
- The development of shale gas would cause a cumulative increase in the European GDP of between 1,600 G€ and 3,400G€ over the period

13. J. Williams, M. Winter, P. Summerton & S. Billington (2013), "Macroeconomic effects of european shale gas production" POYRY – Phase I: gas and power market impacts - Phase II: Employment impact". http://www.poyry.co.uk/sites/poyry.co.uk/files/public_report_ogp__v5_0.pdf, Novembre 2013

2020 – 2050, i.e. an average contribution to annual growth of between 0.3 and 0.6%.

- By 2050, shale gas development would generate between 0.6 and 1 million jobs (direct, indirect and induced).

... the geological, economic and social barriers must be removed

Making this dream a reality is conditioned by knocking down three barriers.

The first is geological. Only when a significant number of pilot projects, each including a few dozen wells, have been launched, will it be possible to estimate the actual state of the resources borne by European source rocks, and their capacity to produce gas once they have been fractured. Are they better, the same, or worse than their US counterparts?

The second is economic. The viability of this kind of project will depend essentially on our capacity to drill and fracture wells at acceptable costs.

To obtain a profitability rate of over 15% considering a discounting rate of 10% and a gas price of US$ 10/MBTU (i.e. approximately US$ 60/boe), wells must be drilled and operated at integrated cost (i.e. incorporating all capital expenses – CAPEX, and operating expenses – OPEX) of somewhere between 7 (for 2 Bcf/well of ultimate reserves) and 16.5 MUS$ (for 5 Bcf/well of ultimate reserves). A similar well in the Barnett shales (3.5 Bcf of associated reserves) should be drilled for approximately 13 MUS$[14].

Since oil services are currently not competitive enough in Europe (few drilling rigs, almost no fracturing fleets, monopoly of a few major international Service Companies), the minimum integrated cost of a well can be estimated at around 20 million dollars. Unless very high potential source rocks are found, the market as it stands would not be conducive to the cost-effective production of European shale gas. Making such developments profitable will require a significant reduction in the cost of wells, in particular by encouraging the emergence of a local competitive

14. All these figures are given before tax.

oil services market, especially in the drilling and fracking sectors. Yet this requires first and foremost determined political backing including a national exploration plan, and possibly a development plan.

The third covers all the societal and cultural issues. In North America, the population has been used to living near drilling rigs, fracturing equipment and production facilities for decades unlike the population of the "old continent", which has barely any oil culture. In a densely populated, urban Europe, hydraulic fracturing, water supply, microseisms and surface impact are seen as a battery of threats by certain stakeholders and to change this perception will take more than scientific reasoning. Winning over public opinion requires pedagogy, transparency, appropriate communication campaigns and commitment regarding local communities[15] with an in-depth review of certain regulations. The game must be win/win so that there is something in it for the majority of stakeholders in the long term. Mining rules which are essentially favorable to the national governments of all Member States, is obviously not a factor that will

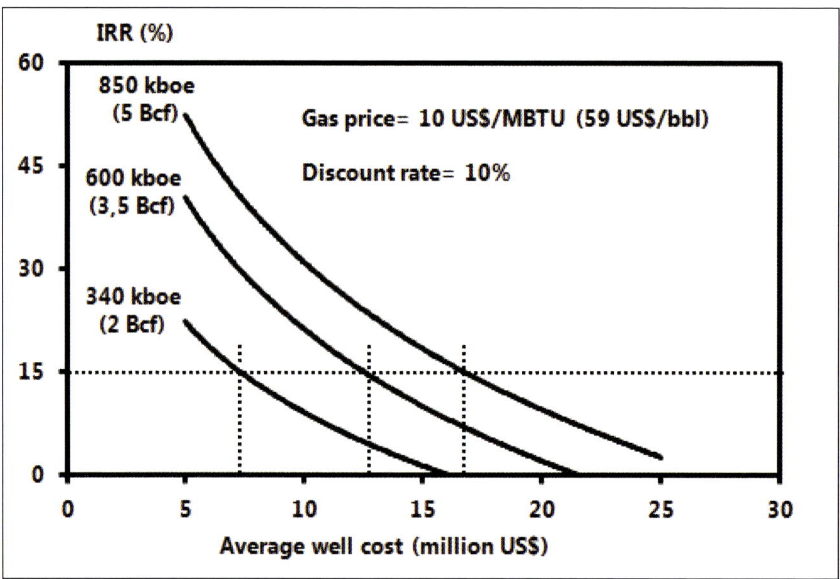

Figure 6 – To obtain a profitability rate of at least 15%, wells must be drilled for 13 MUS$. (before taxes and for a gas price of US$ 10/MBTU)

15. D. Messina (2013), « Shale insights » Shale World May 2013.

encourage local authorities and local communities to be proactive. An improvement in the way in which income is distributed, in particular to local communities, would be a real step in the right direction.

From prohibition to moratorium and from hesitation to contradiction, Europe risks missing out on the energy revolution and ending up as its first victim. In the next few years, the European markets, already strongly disadvantaged by the decrease in demand in the wake of the debt crisis, risk seeing their competitivity handicap increase relative to their American rivals, who benefit from an energy that is two to three times cheaper. The newfound energy independency in the United States is proof that decline is not a foregone conclusion.

Appendix I

The King Hubbert model. Oil and gas peaks excluding shale oil and gas

Using the logistics law of the Belgian mathematician Verhulst[1] designed to model population growth, the famous American geophysicist King Hubbert put forward at the end of the 1950s a predictive model to correlate the production curve to ultimate recovery[2]. Production $P(t)$ over time t is calculated using the equation below:

$$P(t) = \frac{2P_m}{\{1 + \cosh[-b(t - t_m)]\}}$$

where P_m and t_m are respectively the production and the time at peak level, b a dimensionless constant and U ultimate recovery, such that:

$$U = \int_{t=0}^{\infty} P(t)\, dt$$

The relationship between the production peak and ultimate recovery can then be easily calculated:

$$P_m = \frac{Ub}{4}$$

1. http://www.statelem.com/modele_de_verhulst.php
2. http://www.hubbertpeak.com/laherrere/multihub.htm

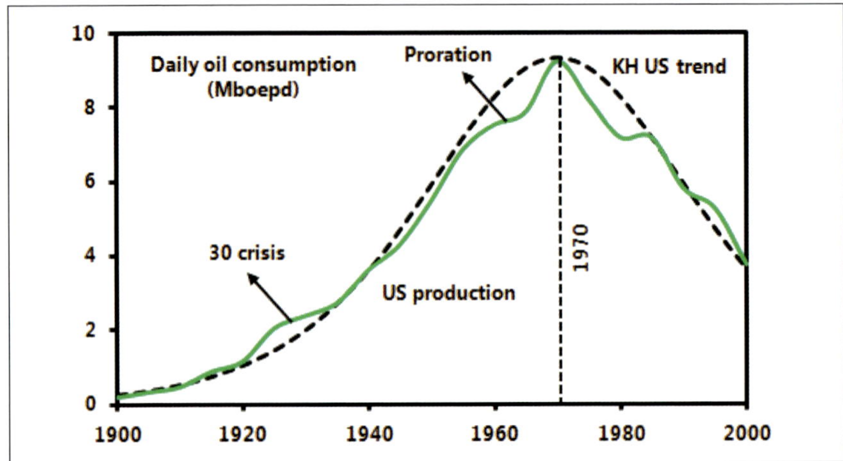

Figure 1 – The King Hubbert model was a perfect simulation of the American oil peak in 1970.

By using this model with the values *U = 184 Gboe* et *b = 0,07*, King Hubbert was able to predict in 1956[3] that the American oil production of the 48 states (excluding the Gulf of Mexico and Alaska) would reach a peak of 9.4 Mboe in 1970. Ridiculed at the time, his analysis later proved to be amazingly accurate (**Figure 2**) even in spite of a number of cyclic events such as the economic crisis of 1929 or proration in the 1960s. Through his analysis, King Hubert also showed that there was a delay of 35 years between the maximum discovery peak (dating from 1935) and the production peak.

Unlike the American production curve which has a bell shape, the world production curve has two very clear growth periods. The first corresponds to the growth of the golden sixties and the second to the growth of consumption in emerging countries as from the mid 1990s. The two upward trends are separated by a plateau of 10 years. The world oil production curve cannot therefore be reproduced using a single model.

If we take the 2012 ultimate reserves (excluding unconventional resources) equal to 2,900 Gbbls (1,250 Gbbls of reserves produced and 1,650 Gbbls of remaining reserves) the world production[4] curve can be

3. King Hubbert (1956), "Nuclear Energy and the Fossil Fuels", Publication Number 95, American Petroleum Institute. http://www.hubbertpeak.com/hubbert/1956/1956.pdf

4. This is actually a world consumption curve.

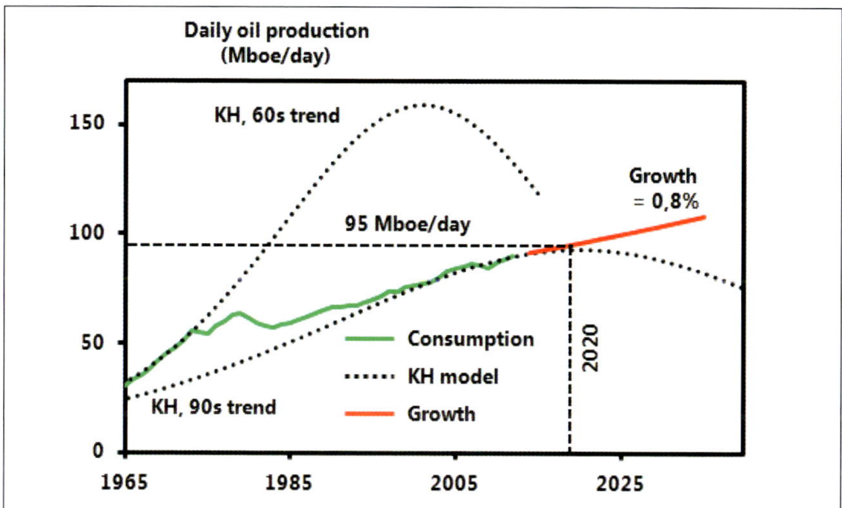

Figure 2 – King Hubbert's model shows that the production trend of the golden sixties would lead to a flat oil peak in 2020

simulated using two different models (**Figure 2**). The first corresponds to the 1960s trend. If this growth in production (around 6% per year) had continued, then oil production would have reached a peak of... 160 Mboe/day in 2000, an absurd figure that could never have been reached. This result demonstrates the lack of realism in the medium term of the golden sixties growth model, artificially stimulated by very cheap energy. The trend of the 1990s is much more realistic, and leads to a flat oil peak at 95 Mboe/day in 2020. According to the growth in demand (0.8%/year) predicted by the International Energy Agency, the oil demand would exceed supply just before the 2020 oil peak. This flat oil peak at 95 Mboe/day in 2020 was confirmed between 2006 and 2010 by numerous experts[5] and also by certain CEOs of large oil companies, like C. de Margerie[6] from Total, who was the first to declare that *"100 Mboe/day is an optimistic case. We will be able to produce 95 Mboe/day but nothing more"*. As for James Mulva[7], CEO Conoco-Phillips, he confirmed: *"the demand will increase, but will be limited by production. I don't think we are going to see the supply going over 100 Mboe/day"*.

5. http://peakgeneration.blogspot.fr/2010/03/world-energy-briefing-hears-of-peak-oil.html
6. Financial Times.
7. http://www.resilience.org/stories/2007-11-10/peak-oil-bp-conoco-and-iea-all-say-its-there

The application of the King Hubbert model to simulate world gas production considering 2,000 Gboe as the ultimate conventional gas reserves, is presented in **Figure 3**. Unlike the oil production curve, gas production grows steadily and can be reproduced easily by a single model. As shown in **Figure 3**, King Hubbert's graph shows a gas peak in 2027 at 71 Mboe/day in line with ASPO[8] predictions. With the 1.8% growth per year in consumption expected by the International Energy Agency, demand would then exceed supply in 2022, i.e. just two years after the oil peak.

Remember that all these results applied before the emergence of unconventional resources.

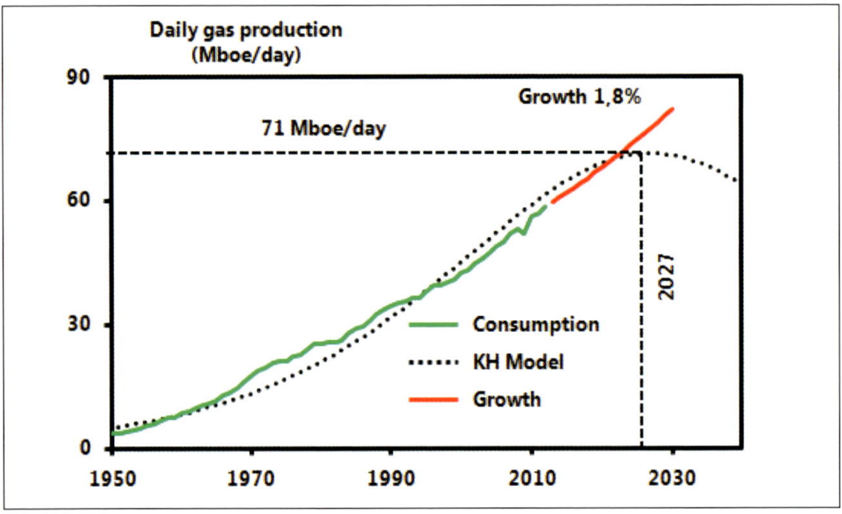

Figure 3 – Application of the King Hubbert model to world gas production.

8. http://aspofrance.viabloga.com/files/JL_veryshort30May2013.pdf

Appendix II

The periodic table of the elements

The periodic table of the elements, also called Mendeleev's table, is a genius classification of atomic elements. It is attributed to the Russian scientist Dimitri Mendeleev (1834-1907). In 1869, he arranged the only known atoms at that time in order according to their atomic mass, which revealed a periodicity in the chemical properties and enabled him to describe certain elements that had not yet been discovered. Each square corresponds to a single element. The number of the square or atomic number corresponds to the number of protons in the nucleus.

The table underwent a series of modifications until it became the periodic table (**Figure 1**) and has become a universal reference and includes 103 elements. The elements in the penultimate line (lanthanides) are also called rare earth elements and are used in particular in the manufacture of photovoltaic cells. The last line (actinides) contains the heaviest elements which are also radioactive, including Thorium (element 90), Uranium (element 92) and Plutonium (element 94) used in the reactors of nuclear power stations.

The four elements of life are element 1 (hydrogen), and elements 6 (carbon), 7 (nitrogen) and 8 (oxygen).

Alkali metals
Alkaline earth metal
Transition metals
Lanthanides
Actinides
Other metals
Nonmetals
Halogen element
Noble gas

1	2	3	4	5	6	7	8	9	10	11	12	13	14	15	16	17	18
1 H (Hydrogen)													6 (Carbon)	7 (Nitogen)	8 (Oxygen)		2 He
3 Li	4 Be											5 B	6 C	7 N	8 O	9 F	10 Ne
11 Na	12 Mg											13 Al	14 Si	15 P	16 S	17 Cl	18 Ar
19 K	20 Ca	21 Sc	22 Ti	23 V	24 Cr	25 Mn	26 Fe	27 Co	28 Ni	29 Cu	30 Zn	31 Ga	32 Ge	33 As	34 Se	35 Br	36 Kr
37 Rb	38 Sr	39 Y	40 Zr	41 Nb	42 Mo	43 Tc	44 Ru	45 Rh	46 Pd	47 Ag	48 Cd	49 In	50 Sn	51 Sb	52 Te	53 I	54 Xe
55 Cs	56 Ba		72 Hf	73 Ta	74 W	75 Re	76 Os	77 Ir	78 Pt	79 Au	80 Hg	81 Tl	82 Pb	83 Bi	84 Po	85 At	86 Rn
87 Fr	88 Ra		104 Rf	105 Db	106 Sg	107 Bh	108 Hs	109 Mt	110 Ds	111 Rg	112 Cn	113 Uut	114 Fl	115 Uup	116 Lv	117 Uus	118 Uuo
		57 La	58 Ce	59 Pr	60 Nd	61 Pm	62 Sm	63 Eu	64 Gd	65 Tb	66 Dy	67 Ho	68 Er	69 Tm	70 Yb	71 Lu	
		89 Ac	90 Th	91 Pa	92 U	93 Np	94 Pu	95 Am	96 Cm	97 Bk	98 Cf	99 Es	100 Fm	101 Md	102 No	103 Lr	

Figure 1 – The periodic table of the elements. The main elements of living matter are 1 (hydrogen), 6 (carbon), 7 (nitrogen) and 8 (oxygen).

Appendix III

Darcy's law

The concept of the permeability of a porous medium was discovered in the XIXth century by the French engineer Henry Darcy (1804–1858). He was born in Dijon, where a square and a fountain now bear his name.

To demonstrate the concept, Darcy created a very simple experiment which entailed forcing the flow of a fluid (water in this case) through a steel tube filled with compacted sand (**Figure 2**). By maintaining a constant hydraulic load (difference between the height of the water columns at the tube input and output), Darcy showed that the flow rate

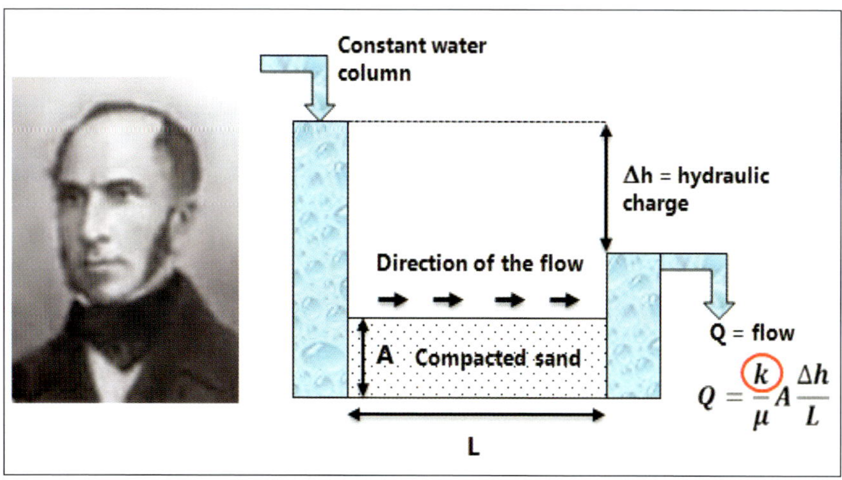

Figure 1 – Henry Darcy (1804 – 1858)
Experiment used by Darcy to demonstrate his law.

of the water running through the sand decreases as the grains become finer and more compact. He translated his experiment into an equation[1] in which he introduced a coefficient *k* (called the permeability coefficient, or simply permeability) which represents the aptitude of a porous medium to let a fluid run through it. The higher the value of *k*, the faster the fluid flows through the porous medium. He also showed that the more viscous the fluid (oil instead of water), the slower it flows through the porous medium. The unit of permeability of a rock is called the Darcy.

- 1 Darcy is a high permeability; that of an excellent reservoir.
- 1 milliDarcy (0.001 Darcy) is the permeability of a poor reservoir.
- Source rocks have permeabilities of between 20 and 500 nanoDarcies (a nanoDarcy is a billionth of a Darcy).

1. Known as Darcy's law

Appendix IV

The porosity of a rock

A clear distinction must be made between the dynamic concept of permeability and the static notion of porosity. The porosity of a rock is the percentage of micro-spaces (called pores – shown in blue in **Figure 1**) it contains, that is the ratio between the volume of spaces and the total volume. The porosity of a standard rock such as sandstone or limestone is about 20%. The size of pores varies from a few microns to a few dozen

Figure 1 – Microphotograph of a rock. The porosity (in blue) is a set of microscopic spaces between the grains. When these are filled by micro-drops of oil or micro-bubbles of gas, an oil or gas reservoir is created.

microns[1]. When these pores are gradually filled with micro-drops of oil and micro-bubbles of gas, an oil reservoir forms, thereby demystifying the widespread notion of an underground oil slick.

1. A micron is a thousandth of a millimeter.

Appendix V

Stokes' law

The free fall (also called "*sedimentation*") of solid particles (sand, glass beads) in a liquid is governed by Stokes' law[1]. It states that the speed of the fall increases as the size (large diameter), and density (specific gravity is higher than that of the fluid) of the particles increases and the viscosity of the fluid decreases.

$$v = \frac{2r^2 g\ \Delta(\rho)}{9\mu}$$

r is the radius of the sphere
v is the free fall velocity
g is the gravity acceleration
μ is the viscosity of the fluid
$\Delta(\rho)$ is the difference betwenn fluid and solid densities

To prove it for yourself, try this simple experiment in your own kitchen. Put two glasses one beside the other. Fill one with water and the other with oil and then sprinkle a pinch of sand over the surface of each glass. Naturally, the grains of sand will settle much more quickly in the water than in the more viscous oil, even though water and oil have a similar specific gravity[2]. For grains and fluids of a given specific gravity, the transportation of solid particles in a fluid will therefore be easier if the fluid is "*thicker*", i.e. if it has a high viscosity.

1. George Gabriel Stokes (1819 –1903), is a British mathematician and physicist. His major contributions to science concern fluid mechanics, optics and geodesy.
2. Oil is slightly less dense than water. To prove it for yourself, pour oil into water: the lighter oil "*floats*".

Appendix VI

Seismic acquisition

Seismic acquisition consists in generating a sound wave (e.g. by a small explosion - **Figure 1**) at the surface. When the wave propagates in the subsurface layers and meets an interface separating two geological layers with different properties, part of the acoustic energy is reflected back to the surface where its amplitude is registered according to time. The arrival times of the reflected waves therefore help locate the different interfaces and deduce an indirect image of the subsurface. This technique was applied for the first time in 1936 in the Ten field near Bakersfield,

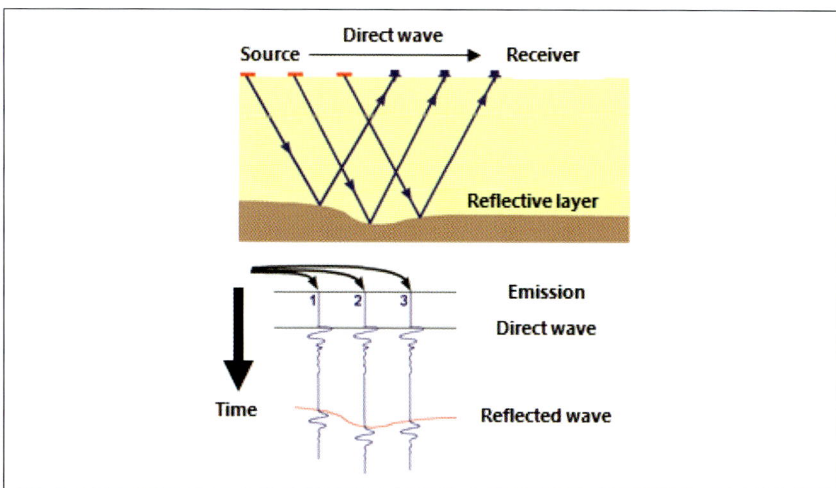

Figure 1 – Basic principle of seismic acquisition

California, and its generalization led to the discovery in the same period of about fifteen giant fields, making the Middle East the world oil Eldorado.

Appendix VII

Project economics

Calculating the profitability of an industrial project sets the capital expenses (investments and operating costs) against the income generated which in the case of an oil or gas project will depend on the production profile (rate at which the field produces gas and oil), the discount rate, (including in particular the rates of the bank from which the operating consortium is borrowing), the oil and gas prices and the contract that links the operating consortium with the producer country, and which includes in particular all the data concerning taxation.

The two main indicators used to evaluate the cost effectiveness of a project are the **Net Present Value** (or NPV) and the **Internal Rate of Return** (or IRR).

The NPV is the additional present income of an investment compared with the capital invested. In theory, a positive NPV indicates that the project can be undertaken. The NPV is calculated using the equation below:
where :

$$NPV = \sum_{p=0}^{p=N} CF(1 + t)^{-p} - I$$

- *CF* is the gross operating profit (reduced by the initial cash outlay),
- *t* is the discount rate,

- p is the annual installment number.
- N the total number of annual installments (total duration of the project).

The NPV remains a provisional evaluation tool based on expenditure/income information that is often difficult to predict. For an oil and/or gas project, all the variables that affect the NPV are uncertain:

- The price of raw materials (price of oil and gas) can fluctuate considerably over short periods of time and depends in particular on unpredictable geopolitical events.
- The production and ultimate reserves per well, which can only be refined over time and as more knowledge of the field is acquired.
- The investment profiles, which in theory should follow a set trend if the project is well defined, have over the last few years followed a high-inflation trend, influenced particularly by steel prices, exchange rates, the increasing demand for local content and the increase in the cost of services.
- The discount rate is also difficult to predict and can have a negative impact on decisions if it is excessively high.

The IRR is the discount rate which cancels out the Net Present Value of a series of cash flows (negative when referring to investments, positive when referring to income). For a project to be viable, its IRR must be sufficiently higher than the discount rate, which corresponds to a positive NPV. The IRR is calculated as follows:

$$NPV = 0 = \sum_{p=1}^{p=N} \frac{FT_p}{(1 + IRR)^p} - I$$

where:

- FT_p is the amount of the pth cash flow,
- N is the number of cash flows (excluding investment),
- I is the initial investment (at date 0),
- IRR is the desired Internal Rate of Return.

Appendix VIII

The water cycle

Considered in ancient times as one of the four fundamental elements of the Universe (with air, fire and earth), water has always fascinated humankind. Over 70%[1] of the Earth's surface is covered in water which has been travelling between the earth and the sky for over three billion years, in its three physical states: liquid of course, but also solid (glaciers, snow) and water vapor (clouds). The water world (or hydrosphere) is made up of two types of systems: those which carry water (rivers, streams that carry water in liquid form, atmosphere in its gaseous form) and those which accumulate water (glaciers, lakes, groundwater layers and oceans). Each accumulator has its own specific dynamics. The time taken to fill and renew them varies from a few years in the case of freshwater lakes, to a few thousand years for oceans and glaciers.

The water cycle can be briefly summarized as follows (**Figure 1**): under the effect of the sun's rays, the water evaporates and enters the atmosphere[2] where it cools, condenses and then falls back to earth as precipitation (rain or snow), then runs off into streams and rivers or infiltrates the subsurface layers and feeds the underground aquifers.

1. Of which less than 3% is fresh water.
2. The quantity of water that evaporates every day from the oceans is estimated at 1,000 km^3.

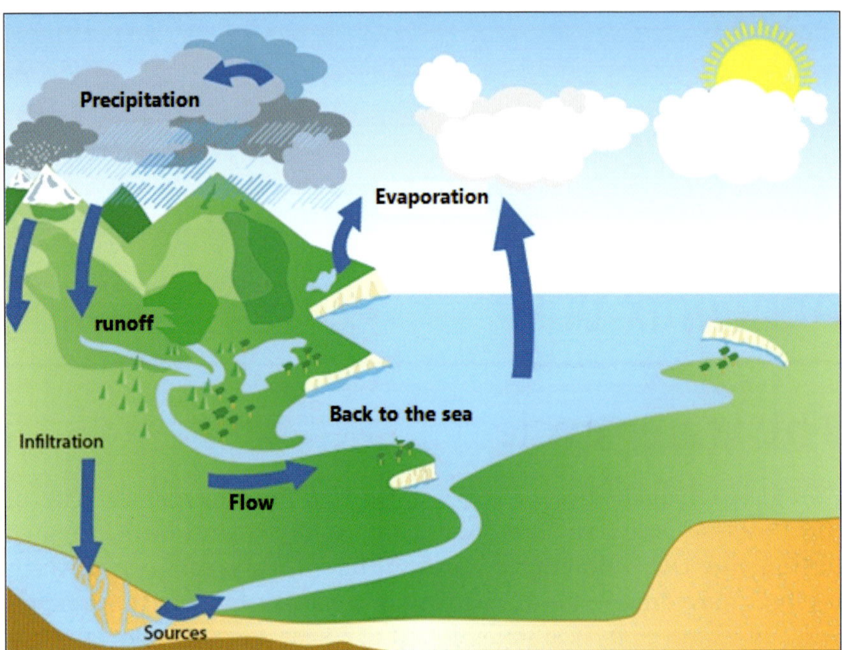

Figure 1 – Water cycle and circulation model (http://www.cieau.com/tout-sur-l-eau/le-cycle-naturel-de-l-eau)

Appendix IX

Radioactivity and shale gases

Sedimentary rocks and particularly shales contain radioactive minerals in small proportions, adsorbed either by organic matter or by the fine clay particles[1]. They include essentially uranium, thorium, radium, potassium and barium. Percentages of uranium of between 0.01 and 0.02% are extremely common. In certain rocks the percentage of Uranium can reach exceptional values of 0.5%[2]. The groundwater solubilizes a small proportions of radioactive minerals and carries them into the subsurface layers. The dominant (because they are the most soluble[3]) radioactive elements present in groundwater are $Radium^{226}$ and to a lesser extent, Potassium, whereas Uranium and Thorium are barely soluble and almost never present in subsurface aquifers. If we limit our analysis to Ra^{226}, its proportion varies widely from one aquifer to another, depending on the rock it washes through.

For this reason, the subsurface aquifers contain traces of radioactive elements in very variable quantities. The surface Aquifer of the Polish Silurian, which is five times more radioactive than the US Barnett one, nonetheless falls within the world average, which is estimated at 85 kBq[4]/m^3 (**Figure 1**).

1. Smith, A. L. (2011), "First Correlation of NORM with a specific Geological Hypothesis", Society of Petroleum Engineers. SPE 138136.
2. Warner, N.R., Christie, C.A., Jackson, R.B., Vengosh, A. (2013), "Impacts of shale gas wastewater disposal on water quality in Western Pennsylvania", Environmental Science and Technology 47, 11849-11857.
3. Smith, K. P. (1992), "An overview of Naturally Occurring Radioactive Materials (NORM) in the Petroleum Industry", Environmental Assessment and Information Sciences Division, Argonne National Laboratory.
4. The Becquerel (Bq) is the SI unit for the activity of a radioactive isotope and corresponds to one disintegration per second. It is equivalent to an inverse second s–1.

During a hydraulic fracturing operation, the large quantities of water that leach through the rock solubilize small proportions of radioactive minerals (mainly Ra^{226} and potassium) which can be found in the fluid recovered at the surface during well flowback phase. The concentration of radioactive minerals dissolved in recovered water is higher than that naturally present in groundwater. The examples of the Polish Silurian and the Barnett shales in the United States show that the concentration in Ra^{226} is 7 times higher in the recovered waters than in the corresponding

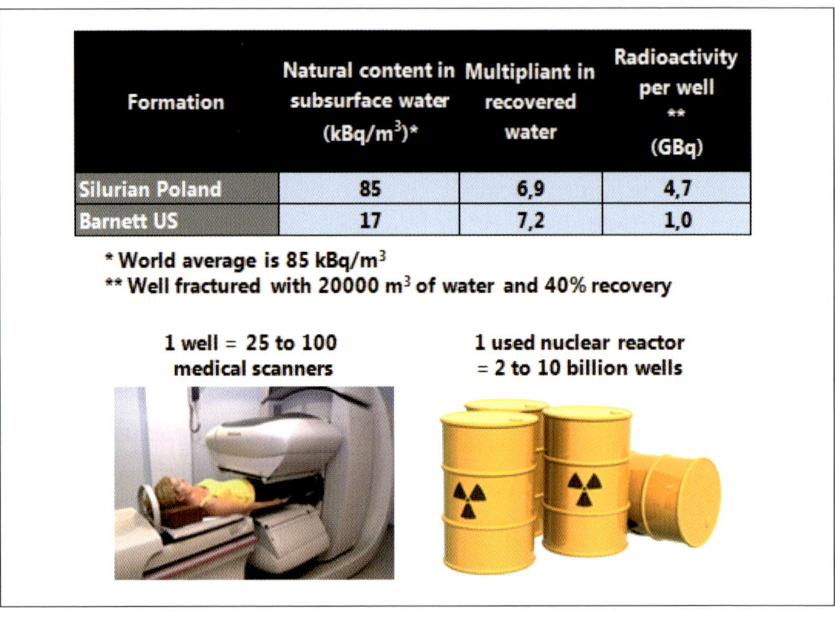

Formation	Natural content in subsurface water (kBq/m^3)*	Multipliant in recovered water	Radioactivity per well ** (GBq)
Silurian Poland	85	6,9	4,7
Barnett US	17	7,2	1,0

* World average is 85 kBq/m^3
** Well fractured with 20000 m^3 of water and 40% recovery

Figure 1 – The groundwater of the Polish Silurian is five times more radioactive than that of the Barnett shales, but is in the world average. The fluid recovered from the Barnett shales and the Silurian is seven times more radioactive than the corresponding natural underground fluids[5]. The fluid recovered from a well of 20,000 m^3 emits the same radioactivity as 25 to 100 X^{ray} scanners. The used fuel of a nuclear reactor emits the same radioactivity as 2 to 10 billion shale wells[6].

5. Almond S., Clancy, S.A., Davies, R.J., Worrall, F. (2014), "The flux of radionuclides in flowback fluid from shale gas exploitation", Centre for Research into Earth Energy Systems (CeREES), Department of Earth Sciences, Durham University, Science Labs, Durham DH1 3LE, UK.
6. A thyroid scintigraphy emits 40MBq whereas the residual radioactivity of a nuclear reactor emits 1013 MBq. http://fr.wikipedia.org/wiki/Becquerel

subsurface aquifers (**Figure 1**). So a well fractured using 20,000 m^3 of water and sending 40% of the injected volume back to the surface will emit between 1 and 5 GBq that is the equivalent of about sixty thyroid scintigraphies. To put these figures in perspective, the best comparison is that of the used fuel from a nuclear reactor, which would emit radioactivity levels equivalent to those of… 2 to 10 billion shale wells.

An even more interesting comparison is that of the emissions generated by different types of energy. **Figure 2** compares emissions/boe for conventional gas, shale gas, coal-fired electricity and nuclear electricity. The results are indisputable: even if the method for producing shale oil and gas emits between 3 and 15 times more than conventional gas, the cycle of a coal-fired power station emits almost 1,000 times more[7]. For a nuclear power station the results are just as clear – emissions are 7,000 times higher.

For humans the health risk of exposure to nuclear radiation is expressed in mSv[8] (milliSievert). In France the advised limit is 1mSv[9] per year per

Formation	Radioactivity per boe* Shale gas	Radioactivity per boe Conventionnal	Radioactivity per boe Nuclear	Radioactivité per boe Coal
Silurian Poland	15	1	7282	959
Barnett Etats Unis	3			

* One supposes an average of 500000 boe (3Bcf) per well

Figure 2 – It is surprising to compare the radioactive emissions of different types of energy. Compared with conventional hydrocarbons (no hydraulic fracturing) shale oil and gas emit 3 to 15 times more radioactivity, but this figure bears no comparison with that of a coal-fired power station which emits almost 1,000 times more and a nuclear power station which generates 7,000 times more emissions[10].

7. The ashes produced by burning coal contain substantial quantities of radioactive minerals.
8. The Sievert (Sv) is used to evaluate the biological impact of radiation on a living organism exposed to radioactivity. The effects vary according to the type of rays and the organs which received the radiation. The most commonly used unit is the millisievert (mSv), which is a thousandth of a Sievert (1/1 000e de Sv).
9. French Public health code, Article R1333-8.
10. Almond S., Clancy, S.A., Davies, R.J., Worrall, F. (2014), "The flux of radionuclides in flowback fluid from shale gas exploitation", Centre for Research into Earth Energy Systems (CeREES), Department of Earth Sciences, 5 Durham University, Science Labs, Durham DH1 3LE, UK.

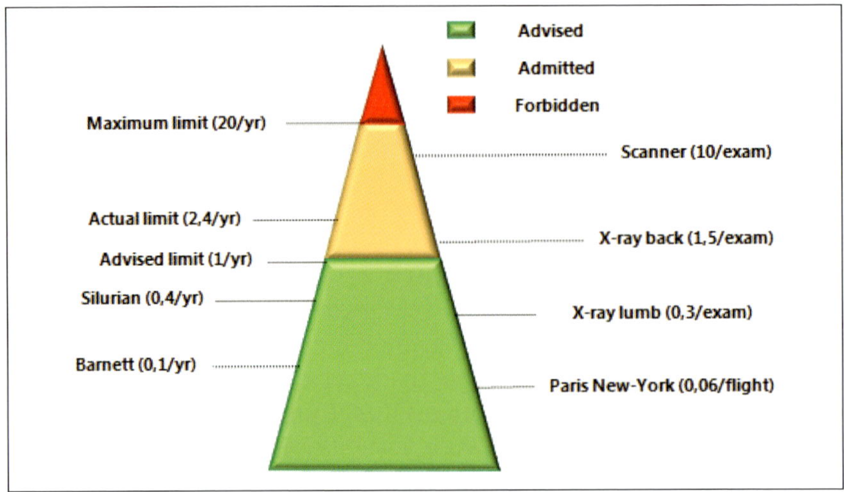

Figure 3 – Radiation exposure (expressed in mSv) pyramid.
In France the advised annual limit is 1mSv per year per person, but in reality a person in France receives an average of 2.4mSv/year. The recovery of the Barnett waters exposes people to approximately 0.1 mSv/year, less than the equivalent of two return journeys from Paris to New York and the Polish Silurian, just slightly more than a lung x-ray and three times less than a back x-ray.

person, but in reality a person in France receives an average of 2.4mSv/year[11]. The recovery of the Barnett waters exposes people to approximately 0.1 mSv/year, less than the equivalent of two return journeys from Paris to New York, and the Polish Silurian, just slightly more than a lung x-ray, three times less than a back x-ray and 25 times less than a medical x-ray scanner (**Figure 3**).

So although the water recovered after hydraulic fracturing effectively contains a greater quantity of radioactive elements than natural groundwater, the resulting radiation is not a health risk. Compared with a coal power cycle and a nuclear power cycle, the production of energy from shale oil and gas contributes an extremely low level of radioactivity.

11. http://www.developpement-durable.gouv.fr/Phenomene-de-radioactivite.html

Appendix X

Some examples of water treatment techniques[1]

Depending on the final destination of the processed water (injection, reuse or discharge), different types of treatments of varying degrees of sophistication can be used. They aim to separate the recovered water from the solids (sand carried along with flow back but also solids leached from the formation) and hydrocarbons (assuming the fluid has been degassed in a green completion) it contains, then eliminate the microparticles in suspension and finally purify the water from salts and dissolved solid particles. Some examples are given below and in **Figure 1**.

Oxidation

This technique extracts from the recovery fluid, the organic matter and certain metals by triggering a chemical oxidation reaction using oxidizing elements such as ozone or oxygenated water. The reaction creates a precipitate which can then be removed from the solution.

Separation

The solids in suspension and the oil are separated from the recovered fluid under the effect of gravity. Gravity separators may simply be closed decanting barrels (static separators). Hydrocyclones operate dynamic

1. P. Quintano *et al* (2014), Recovered Water Management Study in Shale Wells, ERM project 0243719 ordered by International Association of Oil Producers (OGP – http://www.ogp.org.uk/).

separation under the effect of centrifugal force. They are highly efficient at separating water from oil, but do not extract the solids.

Filtration and nanofiltration

Standard filtration extracts large solid particles at atmospheric pressure, such as sand and certain deposits, using simple filters (screens, sand beds or Diatoms which retain the particles and let the fluid pass through). Fine or very fine particles can be extracted from the recovered fluid by microfiltration or even nanofiltration. For nano-filters, porous membranes are used, in which the pores can be as small as a few billionths of a millimeter. A lot of energy is required for nanofiltration as a very high pressure is needed to push the fluid through the nano-filter. For this reason, nanofiltration requires the fluid to be filtered beforehand and then the filtering process is carried out in progressive stages.

Flotation

Flotation involves introducing bubbles of air or gas at the bottom of a closed, airtight tank. As they rise, the gas bubbles cling to oil droplets or solid particles and bring them back up to the surface. The flotation residue is then recovered by skimming.

Distillation

To extract the dissolved solids and salts, the fluid needs to be heated until it boils. The vapor is then condensed, leaving a small volume of brine with a very high salt content. Current technologies are able to treat solid and salt concentrations of up to 40 milligrams/l (i.e. 40 parts per million).

Crystallization

Crystallization consists in eliminating all the water (chiefly by evaporation) to produce a purely solid waste product. Depending on the number of evaporation stages, this technique is able to eliminate up to 95% of the water.

The finer the separation required, the more sophisticated the technique will need to be, the more expensive the cost and the greater the quantity of energy required. So standard separation and filtration will be low cost treatments, oxidation and flotation medium-range treatments and crystallization and distillation the most expensive and the most demanding in terms of energy (**Figure 1**).

Figure 1 – Some examples of separation. From the top left: 1. Activated carbon filters - 2. Precipitation filter press - 3. Microfiltration - 4. Flotation 5. Crystallization - 6. Distillation

In addition to the treatments for separating solids and hydrocarbons and the elimination of dissolved salts and solid particles, the treatment must also remove the bacterial flora, in particular if the wastewater is to be reused, as bacteria are extremely corrosive organisms. Three kinds of treatment are available on the market (**Figure 2**):

- Chlorinated disinfectants similar to bleach.
- Ultraviolet rays that destroy the bacteria.
- Ozone infusion

Figure 2 – Example of a biocide injection unit.

Appendix XI

Greenhouse gases

Some of the sun's incident rays traveling to Earth pass through the atmosphere, while others are reflected directly back into space. Heated by the incident rays, the Earth's surface radiates the heat into space essentially in the infrared spectrum. Nonetheless, only part of this radiation passes through the atmosphere. Greenhouse gases (water vapor[1], carbon dioxide, methane, nitrogen dioxide) that naturally occur in the atmosphere in very low quantities absorb the other part and sends it back to the Earth, keeping its surface at a higher temperature than it would be if the GHG weren't there (**Figure 1**). Yet without GHG, life could not exist on Earth.

However, not all greenhouse gases have the same Gas Warming Power (GWP). The latter is the integral of its "instant radiative forcing" $F_g(t)$ which depends on its spectrum of adsorption over the time it remains in the atmosphere T:

$$GWP = \int_O^T F_g(t)\,dt$$

Figure 1 compares the radiative forcing graphs for CO_2 and CH_4 which show that although methane has a higher instant radiative power, its

1. The atmosphere is mainly made of Nitrogen and Oxygen which are not GHG. GHG represent only a small fraction of the atmosphere. An increase in GHG content in the atmosphere does not generate pollution and does not have any direct impact on health on a global scale.

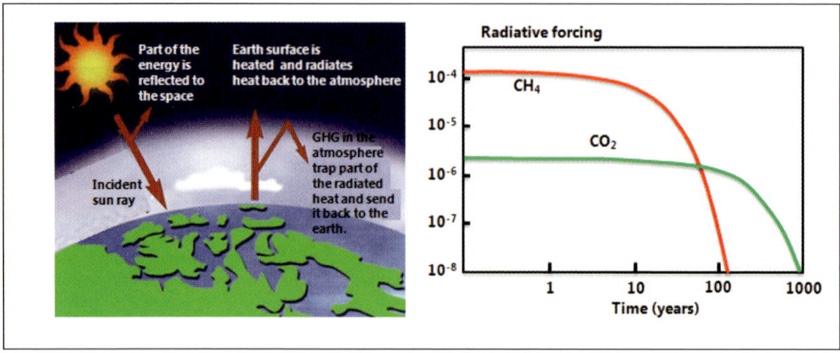

Figure 1 – Greenhouse gas physics. Radiative forcing and lifetime of CO_2 and CH_4. In spite of a much shorter lifespan, the radiative forcing of methane is much higher than that of CO_2. Methane has a GWP 25 times higher than that of carbon dioxide.

lifespan in the atmosphere is much shorter (12 years compared to nearly 100 years for CO_2).

For this reason it is preferable to calculate, over a period of $T=100$ *years*, a relative gas warming power such that:

$$GWP = \frac{\int_O^T F_g(t)\,dt}{\int_O^T F_{CO_2}(t)\,dt}$$

So the relative *GWP* of methane is 25 which means a quantity of methane present in the atmosphere for a century will have an effect 25 times greater than that of CO_2. Certain GHG present in very low quantities in the atmosphere have a very high GWP (298 for nitrogen oxide, 22,800 for sulfur hexafluoride - **Figure 2**).

Increase in GHG owing to human activity

Although GHG are crucial to life on Earth, any excess in the atmosphere increases the temperature at the surface of the planet, which then impacts the climate. It has now been established that the use of fossil fuels (coal, oil, gas) since the industrial revolution has been one of the main contributors to the increase of CO_2 in the atmosphere. The regular measurements taken as of 1958 by Charles David Keeling led to the famous

Gas	Chemical formula	Life span (years)	GWP (100 yrs)
Carbone dioxyde	CO_2	30–95	1
Methane	CH_4	12	25
Nitogen protoxyde	N_2O	120	298
Sulfur hexaflurorure	SF_6	3200	22800

Figure 2 – Gas warming power (GWP) of certain greenhouse gases in the atmosphere.

Keeling Curve or Mauna Loa curve after the name of the Hawaiian resort in which the measurements were taken. In the space of 50 years, the average concentration of CO_2 in the atmosphere therefore increased by 20%, from 320 ppm[2] in 1958 to 390 ppm in 2010 (**Figure 3**).

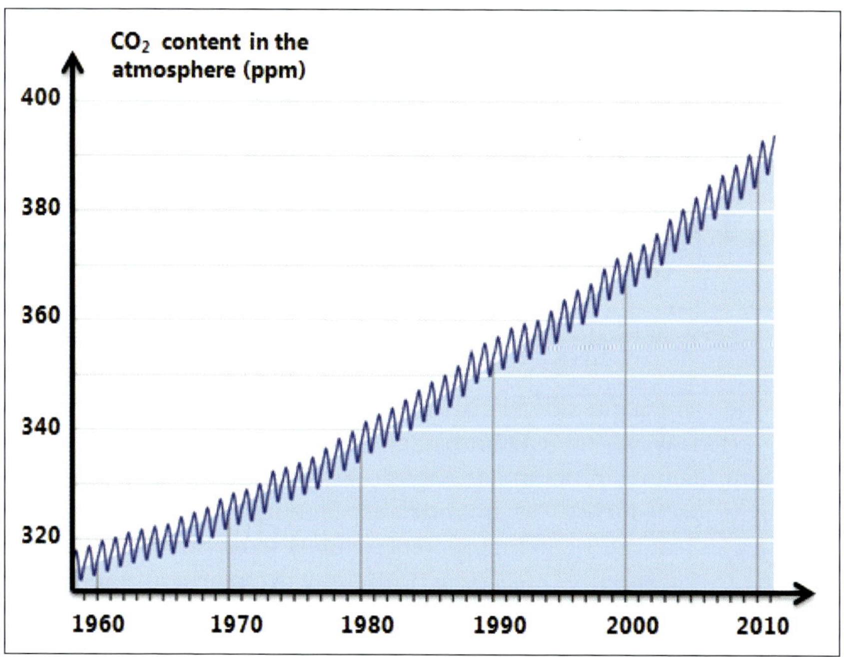

Figure 3 – The Keeling Curve (http://www.climatepedia.org/Keeling-Curve)

2. ppm = parts per million.

Regarding emissions, not all fuels are equal. For an equivalent energy content, the emissivity of coal is 75% higher than that of natural gas, which appears to be far and away the most ecological fuel (**Figure 4**).

To highlight the impact of demographic and economic growth, let's break down emissions as follows[3]:

$$CO_2 = \frac{CO_2}{boe} * \frac{boe}{GDP} * \frac{GDP}{INHAB} * INHAB$$

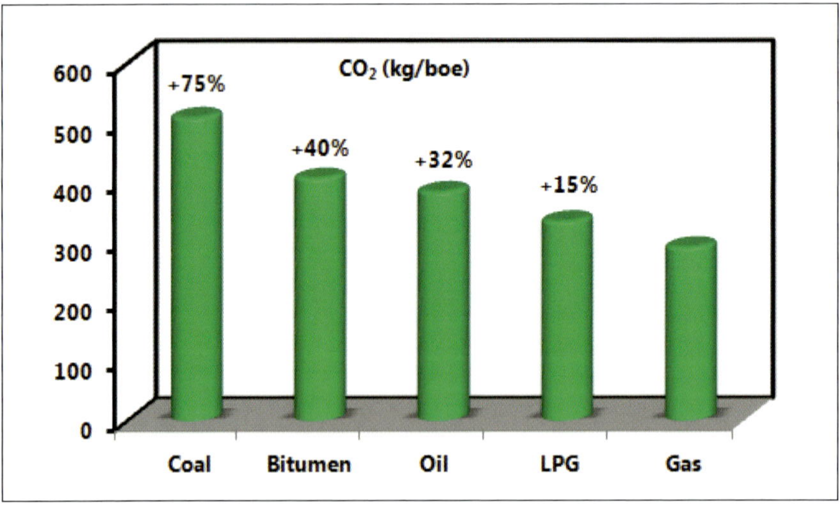

Figure 4 – CO_2 emissions by fuel type

On the right-hand side (1) is the emissivity of a given fuel, (2) the energy intensity, (3) the GDP/inhabitant and (4) the world population. Therefore, demographic and economic growth are not conducive to stabilizing or reducing the quantity of emissions unless the energy intensity is reduced or if high-potential fuels (coal) are progressively replaced by lower-power fuels (gas) or those with none at all (nuclear or renewable).

3. http://manicore.com/documentation/serre/kaya.html

A000023134379